Exit Farming:

Starving the Systems That Farm You

Disclaimer:

This is a true account based on the author's personal experiences and recollections. While every effort has been made to ensure accuracy, some identifying details and timelines may have been changed or condensed to protect privacy or improve clarity. The opinions expressed are solely those of the author.

For Alexys:

You lived this life the same as I did, not watching from somewhere safer, but in it, choosing the uncertain path when certainty was killing us both slowly.

You had the clearer vision when mine was clouded by desperation. You understood which battles were worth fighting and which systems were worth abandoning entirely. You carried burdens I didn't even recognize as weight.

I'm not going to write the kind of dedication that calls you my rock, or my better half, or any of the other generic phrases people use when they can't articulate what actually happened between two people who chose to break everything together.

You're the only one who will truly understand what starvation looks like, not just the absence of food, but the slow dissolution of everything that once felt solid. The particular emptiness of watching a life that should have meaning, drain itself of purpose, one compromise at a time.

I'm sorry for the toll this path demanded of you. I'm sorry for the moments when the weight of what we were building fell heavier on your shoulders. I'm sorry for when your strength had to compensate for my blindness, and your clarity for my confusion.

You mean the farm to me.

I love you.

Me

Table of Contents

Introduction: You Don't Have to Burn Out to Opt Out

Our website says we started farming to feed ourselves. That's partially true. But the full story is a lot messier. We didn't just want fresh eggs or homegrown vegetables. We wanted out. Out of the grind. Out of the debt. Out of a system that rewards burnout and punishes self-reliance.

We were told work hard, go to school, follow the rules, and you'll be set for life. That was the promise. What we got instead was a collapsing economy, unaffordable housing, stagnant wages, broken systems, and a world on fire.

Neither of us started with advantages. After failing out of college my first year, I was power washing the insides of semi-trailers in industrial lots, coming home covered in whatever had been hauled cross-country, earning just enough to keep the lights on with no benefits and no future. Alexys was working double shifts as a CNA, wiping down bodies and dealing with the worst parts of human decay, then following her passion to become a hairstylist but struggling to make rent. We weren't privileged kids rebelling against comfort. We were grinding our way up from the bottom, chasing the promise that hard work actually led somewhere.

We did everything right. Alexys went back to school, earned her degrees all the way through her doctorate, and climbed the corporate ladder rung by bloody rung. She was soon flying to exotic places for high-profile meetings that looked like success but felt like burnout in disguise. I went back to college too, earned a bachelor's and a master's. I clawed my way into federal employment, thinking I'd found security and purpose. We played by every rule they gave us, sacrificed our twenties to their timelines, and watched it slowly consume us anyway. The system that promised to reward our dedication was actually designed to extract everything we had to give, then demand more.

The worst part wasn't the exploitation. It was the realization that there was no finish line. No amount of climbing would ever be enough. The goalposts moved every time we got close, and the prize for winning their game was just a bigger game with higher stakes and less humanity.

For Millennials, the lie wasn't just about prosperity. It was about meaning. We were raised to be everything for everyone, and all at once. Emotionally available, endlessly productive, socially conscious, financially secure. It was never possible.

And now, a lot of us are done. Not with life, but with this version of it.

This book isn't about dropping out and living in a cave. It's about building a different system on the margins of the one that failed you. It's about feeding yourself, literally and figuratively, while starving the ones that tried to feed off you. Exit farming is what we are calling it "internally" from a corporate perspective. Exit farming: Starving the systems designed to farm you.

We don't need a million followers. We don't need a business coach. We don't need to scale. We need to opt out with intention and build a life we don't have to recover from.

This is how we starve the system while feeding our own.

Part I: Seeing the Cage

Chapter 1: The Lie We Were Sold

We were told the same story: follow the path, and it will work out. Go to school, get good grades, land a stable job, buy a house, save for retirement. But by the time Millennials reached adulthood, the path had already collapsed beneath our feet like rotting floorboards in an abandoned house.

The metrics of success were clear and specific, carved into our minds like commandments on stone tablets. I needed my bachelor's degree, then my master's. Alexys continued through her doctorate because that's what serious people did, that's what people who weren't failures did. We needed six-figure incomes. Not one of us, but both of us. That was the baseline for "making it," the minimum entry fee to avoid being discarded by society like yesterday's garbage.

We needed to own a home, obviously, but not just any home. For a brief moment, we owned three at once. We needed the kind of home that projected success, that screamed to the neighbors and relatives that we weren't drowning like everyone else seemed to be. A house that could serve as both shelter and shield against the judgment of a world that had promised us prosperity and delivered poverty disguised as debt.

We were supposed to take expensive vacations to destinations that required long flights and extensive planning. The kind of trips that were more work than they were worth, but that's what successful people did. They suffered for their leisure just like they suffered for their work, posting photos from exotic locations while hemorrhaging money they didn't have to impress people they didn't like.

The retirement target was hammered into us with religious fervor: one million dollars in each of our retirement accounts. Not combined, each. This wasn't really a conversation we had, sitting down at a kitchen table looking at spreadsheets. It was just a fact of life that we were working toward this arbitrary number for some arbitrary reason, like prisoners counting down days to a freedom that would never come. Financial advisors spoke about it like gospel, their eyes gleaming with the fanaticism of true believers selling salvation through suffering. Online calculators confirmed it with mathematical precision. One million each, or you'd be eating cat food in your golden years, assuming you lived long enough to see golden years that weren't tarnished black by poverty and regret.

Our credit scores needed to be in the 800s. We monitored them obsessively, checking multiple times per month for any fluctuation like cancer patients monitoring test results. Once you hit 800, it becomes like a drug. You want to see how high you can get, chasing a perfect score that promises perfect security in a system designed to keep you perpetually insecure. My highest was 847, and I don't care anymore. Most people don't know credit scores are relatively new. They were invented in 1989, 5 years after I was born. They continue to morph and change, and I suspect it's intentional. Credit scores are a moving target designed to keep you chasing something you can never fully control, like a carrot dangled in front of a donkey that's being led to slaughter.

The rules kept shifting like quicksand beneath our feet. Carry this much debt to income, but not too much. Don't take out this type of loan but do take out that type. Keep your utilization below 30%, but not at zero because that looks suspicious to the algorithm gods who determine your worthiness. Open new accounts to improve your mix, but don't open too many because that looks desperate. Pay everything on time but keep some debt because paying everything off completely looks risky to lenders who profit from your perpetual servitude. The system was designed to keep you guessing, always one step away from the perfect score that would unlock the perfect life. That life will always be just out of reach, like trying to grab smoke with bare hands.

We looked into the FIRE (Financial Independence, Retire Early) movement. We thought FIRE might be our escape route, our tunnel out of the prison we'd built around ourselves. But it seemed absurd to bet your entire future on other people playing with your money in markets that are essentially legalized gambling. They call them speculators for a reason, right? The whole thing was built on assumptions that the markets would take care of people, that volatility somehow smoothed out over time like a sea that only appeared calm from a distance, that nothing bad would ever happen to disrupt the perfect mathematical models that had already failed every previous generation that trusted them.

We entered the workforce into a recession, loaded with debt for degrees that no longer guaranteed anything except a lifetime of payments. We watched housing costs explode like bombs going off in slow motion, wages freeze like bodies in a morgue, healthcare become a luxury reserved for the wealthy while the rest of us died slowly from preventable diseases. Jobs disappeared or became gig work with no benefits, leaving entire generations to compete like gladiators for the privilege of corporate servitude. Then we were told we weren't trying hard enough, that our failure was a character flaw rather than a systematic slaughter.

We were sold stability and handed burnout like a poison pill disguised as medicine.

The Federal Meat Grinder

My last federal job was as a GS-14, twelve years into my career, twelve years of my life fed into a machine that chewed up competent people and spat out broken shells. For context, GS-14 is the second-highest pay grade in the federal General Schedule system, typically reserved for senior advisors, top-tier specialists (like myself), and those just below executive ranks. It's the level where you're supposed to have made it. The promise was that government work meant stability, decent benefits, and steady progression up a ladder that led somewhere worth going. The reality was a toxic environment where politics mattered more than competence and seniority meant nothing except that you'd been suffering longer than the newer victims.

There was a moment. Small on the surface, easy to dismiss. But it laid the groundwork for something larger. I was pulled aside by my supervisor and told, without irony, to stroke a coworker's ego. That was the exact phrasing. Not support. Not mentor. Stroke. This coworker had less than three years in. I was twelve years into my federal career.

What does that even mean, stroke his ego? Tell him he's brave for sending an email? Gasp at his PowerPoint transitions? Pretend his blank stare during meetings is deep strategic thinking?

I didn't do it. And that's when the punishment came.

I was put out to pasture. Three weeks of nothing. No projects. No meetings. No explanation. I didn't exist. Not to anyone. Just silence. The quiet machinery of punishment doing what it does best. Marking someone for slow removal. A soft firing. Not official, not documented. Just a silent warning. Comply, or disappear.

"Put out to pasture." It's a real phrase, and it's like corporate solitary confinement, but with nicer clothes. This is how they create bad performance reviews in the federal system, how they manufacture failure in people who refuse to fail on their own. They starve you of work, then document that you're not productive. It's a deliberate process designed to push out anyone who won't play along with the dysfunction, anyone who still has enough soul left to recognize corruption when they see it. But I take incredible documentation and am great at finding work or hopping on projects to help others. I survived the attempted sabotage like a wounded animal limping away from a trap, but the message was clear: conform or suffer, bend the knee or break.

The drive to work became a daily exercise in psychological preparation for battle, like a soldier suiting up for a war that never ended. Mentally gearing up to deal with incompetent managers who'd failed upward, backstabbing colleagues who'd traded their integrity for job security, and work that served no meaningful purpose except to keep the machine running and the money flowing upward to people who produced nothing of value.

The commute home was decompression time, trying to shake off the toxicity like a dog shaking off fleas before walking through our front door. But even home didn't feel like refuge anymore. The poison followed us home, seeping into our evenings like carbon monoxide, invisible, odorless, and slowly killing everything it touched.

The Dream Home That Became a Nightmare

The feeling on the drive to close on our second house was excitement, whether we were pretending or not, neither of us will ever know for sure, and maybe that's the most terrifying part. You create emotions sometimes when you know you're supposed to feel a certain way, when society has programmed you to celebrate your own imprisonment. This was supposed to be our dream home, the culmination of years of hard work and smart financial decisions, but it felt more like signing our own death warrant with a smile.

The closing itself was routine paperwork and signatures, but underneath the mundane process was this sense that we were locking ourselves into something permanent. More debt, more responsibility, more chains wrapped around our necks disguised as keys to success.

The first night in our new home, it didn't feel like ours, and it never did. We walked through rooms that looked like they belonged to someone else, someone we were pretending to be, someone we'd been told we should become. Everything was tasteful and expensive and completely soulless, like walking through a museum of other people's lives, or a funeral home decorated to make death look comfortable.

I remember sitting in the basement, working from home in what should have been my perfect office space. Looking around, thinking: this is really it, huh? That's it. Buy all this shit and pay for it dearly for the rest of your life, and then die. Work until you're too old to enjoy anything, then die surrounded by expensive things that never brought you happiness. My father used to predictably tell me "life sucks and then you die..." and coming from a man who made life suck for everyone around him, it felt less like advice and more like a confession.

The weighted blanket feeling wasn't metaphorical. It was a physical sensation, like the house itself was pressing down on us, crushing our lungs slowly and methodically. The mortgage payments, the insurance, the maintenance, the HOA fees, the property taxes. All of it added up to this crushing weight that we'd voluntarily placed on our own shoulders like carrying our own coffins while we were still breathing.

We'd sit in our beautiful living room with our expensive furniture, looking out at our manicured yard in our desirable neighborhood, and feel completely disconnected from our own lives. This was supposed to be success.

The Farm Revelation

We started taking drives on weekends, just to get out of the house and away from the feeling of being trapped in our own success like animals in a gilded cage. The first time we drove to what is now our farm was on Thanksgiving weekend. Four hours to see a property we'd found online, just to look at something different, something that didn't feel like a lie wrapped in expensive wrapping paper.

We fell in love that day and never looked back. Not with the property specifically, but with the feeling of being somewhere that felt real, somewhere that didn't require us to pretend to be people we weren't. Rolling hills, open sky, quiet that wasn't broken by traffic or neighbors or the constant hum of suburban anxiety. The silence wasn't empty, it was full of possibility, full of the chance to breathe without choking on other people's expectations. Don't get me wrong, the property was absolutely beautiful, and still is. But now it's a dream come true.

The drive back to our dream home that evening was brutal, like returning to prison after a brief taste of freedom. Every mile closer to our "successful" life felt like descending deeper into hell, each familiar landmark a reminder of how trapped we really were. We didn't talk much on the drive, but we both knew something had shifted. We'd seen what peace looked like, and it wasn't what we'd been sold.

After that, the weekend drives became regular escapes, like brief prison breaks that only made the return more unbearable. We'd load the dogs and cat into the car (yes, the cat loves car rides too) and head west, anywhere that felt more authentic than our manicured subdivision. We were never home because home didn't feel like home, it felt like a stage set for a play we didn't want to be in, a beautiful trap that got smaller every day.

The Generational Divide

The generational disconnect was brutal and personal, like watching your parents speak a language you'd never learned while insisting you were the one who couldn't communicate. My parents are boomers who were able to purchase multiple homes without college educations or six-figure incomes. They had cars and projected a luxurious lifestyle on regular working-class incomes, living in a world where loyalty was rewarded and hard work actually led to security instead of exploitation.

My parents received a loan on land from their parents and then built a big home on a bluff next to doctors and financial advisers, the outward projection of wealth and class that they thought mattered, that they believed would protect them from the economic violence that was already destroying their children's futures.

But they couldn't recognize that we were born at a different time with no jobs, recessions that never really ended, crazy inflation that made their achievements impossible to replicate, and completely different rules written by people who wanted to extract everything from us while giving back nothing. They don't understand that thirty years can pass, and the job market isn't the same as it was when they entered it, that the social contract they'd lived under had been shredded and burned before we ever got a chance to sign it.

When things were hard for me at work, when I'd come home frustrated by the dysfunction and politics, when I was drowning in a system designed to break me, I'd sometimes call my parents looking for support or perspective. The response was always the same: "Take care of your job, kid." Not advice about how to handle difficult situations or toxic colleagues. Not empathy for the impossible position I was in. Just the same reflexive command to keep my head down and be grateful for what I had, like telling someone to appreciate the quality of the rope while they're being hanged.

"Take care of your job, kid." The phrase was hammered into me like nails into a coffin. Keep your head down. Don't rock the boat. Take care of your job because your job will take care of you. Except the job wasn't taking care of me. The job was slowly killing me, drinking my blood while promising me a pension I'd never live long enough to collect, and my parents' advice was to let it happen quietly, to die with dignity like a good soldier.

They couldn't see that the rules had changed, that the game they'd won was rigged against us from the start. That loyalty to employers was no longer rewarded with security but punished with exploitation. That working hard and following the rules no longer guaranteed a comfortable retirement but ensured a lifetime of servitude to people who viewed us as expendable resources. That the path they'd followed successfully was now a trap designed to catch and kill their own children.

The Breaking Point

Before we walked away, we tried to do everything "right," like good little soldiers following orders toward our own destruction. At one point, we were carrying two mortgages, our home and an income property, while juggling two six-figure jobs and expensive car payments. We owned two different sized campers for different purposes, accumulating tools for relaxation that felt more like work than our actual work. We should have felt secure. The truth is, we weren't even stretched financially, on paper, we were fine. But it still felt like we couldn't breathe, like drowning in shallow water while everyone around us insisted we were swimming.

The moment of recognition came right after we closed on our second home, like a final click of handcuffs locking into place. We had turned our first house into an income property, upgraded to what should have been our dream home, and financially we were in good shape. But immediately after signing those closing documents, this dark consumerist cloud started swallowing us up like quicksand. More money, more problems indeed, more money, more chains, more reasons to stay trapped in a life that was killing us slowly and methodically.

Looking at our stuff, the two homes, the two campers, all the things we'd accumulated like barnacles on a sinking ship, it was just an outward reflection of people we weren't with things we didn't want. Our whole life felt foreign, like wearing someone else's skin that didn't fit. Everyone else seemed very comfortable wanting to settle in, but we were constantly leaving, constantly fleeing from the beautiful prison we'd built around ourselves.

We didn't initially admit it to each other, but in less than six months we were ready to liquidate everything, ready to burn it all down before it burned us first. We talk and communicate very well, so once we both felt the walls closing in, the decision came quickly and desperately.

That's the beginning of real freedom. Not chasing comfort, but reclaiming control before the system finishes what it started. Not optimizing our cage, but breaking out before we died inside it.

People like to say we're lazy, that we took the easy way out. But it takes more grit to start from nothing and build your own food, water, heat, and income than it does to sit in a cubicle following orders toward your own slow death. It takes more courage to walk away from everything you've been told defines success than it does to keep grinding away until there's nothing left of you but debt and regret.

Now, when I hear financial advisors or articles pushing that million-dollar retirement number, or whatever arbitrary figure they've pulled straight out of their assholes, I smile grimly as I look down at the rabbit we raised and am eating myself, or the eggs we're having for breakfast that were laid feet from where we sleep. I know that the stock market won't put this food on my table in perpetuity, that those numbers in accounts managed by strangers are just promises written in disappearing ink. Real security isn't a number in an account managed by people who profit from your failure. It's the ability to feed yourself regardless of what happens to everyone else's money, to survive when the system that tried to kill you finally collapses under its own weight.

This book is for the ones who feel the same darkness closing in around them. Not the ones hoping the system will fix itself, it won't, because it's not broken, it's working exactly as designed. But the ones ready to build something better with their own hands before it's too late, the ones willing to burn the rulebook they were given and write their own ending before someone else writes it for them.

Chapter 2: Dependence by Design

It's easy to think the system failed us. But the truth is harder and more terrifying: the system was designed to extract from us like a parasite feeding on its host. It was built to trap us, not support us, to slowly drain our blood while keeping us alive just long enough to be profitable. Every institution, banking, healthcare, housing, education, employment, is built around a single idea: keep people dependent like cattle in a feedlot, even if they're doing everything "right," especially if they're doing everything "right."

For us, it was mortgages, car payments, and jobs. All of it. We had the house, the income property, the careers that looked good on paper. We bought the car that should've made life easier. None of it brought peace. It all brought pressure.

The Bike Ride Revelation

I remember the exact moment I said I wouldn't oppose getting hit by a car. I was riding my bike to work, thinking that maybe if I could make the commute healthier and more environmentally conscious, it would somehow make the job more bearable. All those bike-to-work days and corporate wellness programs are a joke. No amount of biking makes that life palatable.

I mentioned it casually as I was getting ready to leave one morning, like I was commenting on the weather. "I don't plan on killing myself, but if I got hit by a car, I wouldn't oppose it." Just matter-of-fact acceptance that my job was slowly killing me anyway, so what difference would it make if the process sped up?

That's what "taking care of your job, kid" had gotten me. That's where following the rules and keeping my head down had led. A federal position with good benefits and steady pay, and I was fantasizing about traffic accidents as relief.

The bike rides didn't help. The corporate wellness initiatives didn't help. The employee assistance programs didn't help. Because the problem wasn't my attitude or my fitness level or my coping strategies. The problem was a system designed to extract everything from me while giving back just enough to keep me functional.

Midnight Databases and Legal Deadlines

My federal work came with expectations that bled into every hour of every day. I remember being in bed on my laptop at midnight, updating GIS databases and testing functionality for lawyers. Trying to communicate technical things to the least technical people. Emergency deadlines that somehow always became my problem to solve, regardless of the hour or the day of the week.

The job description never mentioned 24/7 availability, but that's what the culture demanded. Not officially, of course. Officially, the government respects work-life balance and employee wellbeing. Unofficially, the people who got ahead were the ones who sacrificed everything for the mission, and the people who set boundaries were quickly marginalized.

General counsel didn't call at reasonable hours because reasonable hours weren't when they felt like working. The urgent requests came during dinner, during weekends, during vacations. Always with the implicit understanding that saying no wasn't really an option if you wanted to keep your job and your security clearance and your carefully built career.

The Snake Oil That Broke Everything

My worst day at that job had nothing to do with databases, deadlines, or even the nonstop political games. It was the day a coworker tried to use Alexys' Multiple Sclerosis as a sales opportunity. Not to offer support. Not to ask how she was doing. To sell me something. Unproven supplements. Miracle cures. Snake oil. Pitched like a product demo in the middle of the workday. He saw her illness not as something human or serious, but as a chance to make a commission.

I reported it. Not just the ethics of it, but the fact that someone was using a federal office to run a personal hustle. Instead of addressing it, my supervisor turned it against me. I was put out to pasture, again. Not because I caused a problem, but because I refused to look the other way while someone tried to profit off the person I love.

I was out to pasture so long that I had to go around my supervisor just to get out. I filed a transfer request directly with division leadership, because everyone already knew how toxic my supervisor was. But nobody was willing to act on how horrible of a human being my supervisor was. That was the culture. Work around the rot, keep your head down, and protect the dysfunction at all costs.

I still remember the day I finally got my transfer. I waited months for no apparent reason. I arrived at my office at 4 a.m. on a Friday. My supervisor and I worked four 10-hour shifts, so Fridays were usually our day off. I packed up everything I owned, left the key on the desk, and never stepped foot in that office again.

It felt like getting paroled from a sentence they spent years pretending I volunteered for. And even then, I had to sneak out like a criminal just to avoid the fallout.

My direct supervisor was notified the following Monday. That's how the official move went, under cover of darkness, quietly, so I wouldn't get caught doing something leadership had already approved. That's how backwards it had become. Even leaving required secrecy. Even doing the right thing felt like breaking the rules.

That's when we understood the truth. The system wasn't broken. It was operating exactly as designed. Not to reward competence, or integrity, or doing what's right. But to maintain control, and push out anyone who dared to expect better.

The Subscription Revelation

The extraction mechanisms were everywhere, some obvious, some sneaky. The big expenses like mortgages and car payments were visible, but the smaller bleeds were almost invisible until you added them up.

We had music streaming services we rarely touched. Every TV streaming service imaginable, just in case we wanted to watch something on each one. Software subscriptions for programs we opened maybe once a month. But that's the game, isn't it? Monthly or yearly billing. No refunds for the time you didn't use.

Each charge felt harmless on its own. Ten dollars here. Fifteen there. Twenty-five for the premium version of something we were convinced we needed.

But when we finally sat down and calculated what we were paying for non-tangible things we didn't own, the number was staggering. Hundreds of dollars every month flowing out of our accounts automatically, for access to things that could disappear at any time. Music libraries that vanished when licensing deals changed. Movies that became unavailable when contracts expired. Software that stopped working if a payment is missed.

We were renting our entire digital lives, paying monthly fees for the privilege of accessing things we'd never actually own. And when we realized this, the response was immediate and final: "That's it. We're canceling everything."

Not a gradual reduction or careful evaluation of which services provided the most value. Just complete rejection of the rental economy that was quietly bleeding us dry. We went from multiple streaming subscriptions to buying CDs. From software subscriptions to one-time purchases of tools we could own permanently. From renting our entertainment to owning our media.

Starting to See the Design

We started realizing we were renting our own time. You had to pay a lot of money to get the most relaxing time off. The system was designed so that even our rest came at a premium.

We were taught to chase credit scores like our lives depended on it. We were told to max out retirement accounts, to keep stable jobs, to invest in things we didn't believe in, and to avoid risk at all costs. And we believed it because that's what was drilled into us. But none of that protected us when things started falling apart.

What kept us stuck wasn't comfort. It was fear. The fear of losing the little ground we had. The fear of starting over. The guilt of letting go of things that took so much work to earn. Fear of losing health insurance. Fear of damaging our credit. Fear of disappointing family. Fear of social judgment. But the longer we stayed, the more we realized: we weren't free. We were just tired.

For me, the burnout was gradual and then very rapid as I attained new jobs with higher grades in the federal government. The people got shittier, the ethics got iffier. The higher I climbed, the worse it became. Being used up and spit out looked like longer hours for the same pay, dealing with supervisors who protected problem employees while targeting conscientious ones, watching good people get pushed out while incompetent people got promoted.

Eventually, the fear started to fade, not because it disappeared, but because we got too angry to care. We saw it for what it was. The debt, the jobs, the insurance, the endless bills, none of it was an accident. It was a design. A system that needs you to believe there's no other way.

It's all coordinated if you follow the blueprint of the American dream. Each step leads to the next level of dependence. College debt leads to job dependence. Job dependence leads to location dependence. Location dependence leads to mortgage dependence.

Mortgage dependence leads to credit dependence. Credit dependence leads to insurance dependence. Insurance dependence leads to more job dependence. The cycle never ends. The grass is never greener, and is most certainly filled with dog shit.

For people who say these systems provide value and convenience, they're absolutely right. We determined that our time together with one another and living an intentional life was more valuable than the convenience they offered.

The design isn't hidden or secretive. It's right there in the marketing. "Build your credit." "Invest for your future." "Protect your assets." "Maximize your benefits." All of it designed to keep you engaged with systems that extract wealth and time and energy while giving back just enough to keep you hoping things will eventually get better.

But they don't get better if you stay in the system. They get tighter. They get heavier. And then one day you're too deep in to move.

Chapter 3: The Real Cost of Convenience

People think convenience is neutral. But it comes with a price: one that most don't notice until it's already stolen something from them, like carbon monoxide poisoning that kills you while you sleep, odorless and invisible until you're already dead.

We don't use DoorDash. We don't stream music. We buy CDs and play them like it's 1999. Not because it's hip, but because it's real. It doesn't spy on us like a digital stalker. It doesn't nudge us into paying for convenience we didn't ask for like a drug dealer offering free samples. We own the music like owning our own thoughts. The artist gets paid fairly instead of being exploited by platforms that treat creativity like a resource to be strip-mined. The record company doesn't disappear our library when they lose a licensing deal, like burning books we'd already paid for.

You don't see the cost of convenience until you try to do something for yourself and realize you don't know how, like discovering you've been slowly paralyzed without noticing. I never learned how to wash my clothes by hand. I had never done dishes by hand until we moved to the farm. There are simple things that machines can't do once they break, and I should be able to do them with little to no lag between: basic human skills that had been surgically removed from my brain to make room for corporate dependencies. I'm not anti-machine or anti-tech, but if things fail, I need to be capable of surviving without them, like knowing how to swim before you get on a boat that might sink.

The First Kill

The first rabbit I processed changed everything, like the first crack in a dam that's been holding back a flood. I never learned how to process my own meat until I butchered a rabbit for the first time, and that day I felt sufficient and successful for the very first time in my life. Not the fake success of paychecks and promotions, but the deep satisfaction of competence, of being able to sustain life with my own hands instead of depending on systems designed to exploit both me and the animals we eat.

Rabbit processing is straightforward once you understand it, like most essential skills that have been hidden from us to keep us dependent. You need a bucket, a hose, a sharp knife, kitchen shears, a rabbit wringer or broomstick, and a gambrel. All the tools necessary for quick, clean, and efficient processing. The setup is simple, the process is ancient, and the result is the freshest meat you'll ever eat: food that hasn't been shipped across continents, pumped full of chemicals, processed in factories that treat animals like manufacturing inputs.

Before I went out to do it, I double and triple checked with Alexys that it wouldn't change her feelings about me, that she wouldn't see me as some kind of monster for taking a life to sustain our lives. Would she see me differently if I killed an animal we'd raised? Would it disturb her to know I was capable of that kind of violence, even necessary violence? She assured me it wouldn't, and it didn't. If anything, it made her proud that we were becoming more self-sufficient, that we were breaking free from the industrial death machine that kills billions of animals in conditions that would give most people nightmares if they ever saw them.

Physically, I didn't feel much during the process itself. I was numb to the emotional aspects because I was focused on being competent and efficient, on ending the animal's life as quickly and humanely as possible. My main concern was for the rabbit: did I kill it quickly enough? What could it feel? Was I minimizing its suffering? The responsibility of taking a life, even for food, wasn't something I took lightly, but it was a responsibility I was finally ready to accept instead of outsourcing to strangers who care nothing about suffering or dignity.

But the moment it was done, something shifted inside me like tectonic plates finding a new position. We had raised this animal from birth, cared for it, fed it, provided for its needs, and now I was processing it into food for our table. The entire cycle was under our control, from beginning to end. No factory farms where animals live in their own waste. No industrial processing plants where workers are treated almost as badly as the animals. No wondering about the conditions or the chemicals or the treatment of the creatures that died to feed us.

The meat went directly into our crock pot, as fresh as meat can possibly be. We'd gone from a living rabbit to dinner preparation in under an hour, a timeline that would have been normal for most of human history, but feels revolutionary now that we've been trained to think food comes from plastic packages.

The Meal That Broke the Spell

That first rabbit dinner was unlike anything I'd ever experienced, like tasting real food for the first time after a lifetime of eating cardboard. It tasted unlike anything I'd ever had: not just because I'd never eaten rabbit before, but because I'd never eaten truly fresh meat, meat that hadn't been killed weeks ago in some distant slaughterhouse and shipped around the country like a corpse making the rounds at funeral homes.

The conversation during dinner was pure envy of ourselves, but the good kind that comes from achieving something you didn't know you were capable of. "I can't believe we're actually eating this right now," but in the best possible way. Not disbelief that we were capable of it, but amazement that we'd actually done it, that we'd completed the entire cycle from raising to processing to eating without depending on anyone else's systems or standards or conscience.

This was what food was supposed to be, what our ancestors would recognize as normal: knowing exactly where your protein came from, how it lived, how it died, what it ate. This was what self-sufficiency actually looked like. Not just growing vegetables in a backyard garden like playing house, but producing protein from animals we'd raised ourselves, taking complete responsibility for our food chain instead of trusting it to industrial systems that prioritize profit over everything else: over animal welfare, over worker safety, over environmental protection, over human health.

Every bite was satisfaction earned through work and intention, every bite a middle finger to the system that wants to keep us dependent and ignorant. Every bite was proof that we didn't need to depend on grocery stores and processing plants. We didn't need to depend on distribution networks that we couldn't control or trust, systems that could disappear tomorrow and leave us starving like most people would be. Like most people will be.

The Theft of Capability

Every machine that saves us time also steals our skills, like a deal with the devil that seems good until you realize what you've traded away. The washing machine. The dishwasher. The streaming algorithm. They're replacing us quietly, one button at a time, until we're completely helpless without them, until we can't survive a power outage or a supply chain disruption or any interruption in the systems we've become addicted to.

I wasn't squeamish about processing the rabbit: I was mad, burning with the kind of rage that comes from realizing you've been lied to your entire life. Mad at a world that convinced me I was incapable of basic human skills. Mad at myself for believing it, for accepting helplessness as normal. Mad at a system that deliberately keeps people dependent by making basic life skills seem impossible or unnecessary, like amputating people's hands and then selling them expensive prosthetics that break down regularly.

How many people know how to process their own meat? How many people know how to wash clothes without a machine? How many people can cook a meal from ingredients they grew themselves? How many people can fix basic problems without calling a service or buying a replacement? The convenience economy hasn't just made us soft: it's made us helpless, and helpless people are profitable customers who need to buy solutions to problems they used to solve themselves.

Now, when I wash dishes by hand, I control exactly how much water I use, like controlling my own life instead of letting machines and systems control it for me. I don't waste it because every drop is precious when you're responsible for it yourself. I don't pay more for it because I'm not paying for the convenience of ignorance. When something breaks, I fix it instead of replacing it, learning more about how things work instead of becoming more dependent on things I don't understand. It all adds up. It's not just about savings: it's about ownership. Of time. Of attention. Of resources. Of capability. Of dignity.

The Blank Stares of the Living Dead

We've had people see the way we live: hauling water, washing dishes by hand, buying physical media, and they give us blank, polite stares and quickly pivot to something else, like looking at an exhibit in a museum that makes them uncomfortable. They gloss over it because they don't get it, and they don't want to get it because getting it would mean acknowledging how helpless they've become. It makes them uncomfortable, like showing a mirror to someone who's forgotten what they look like.

It doesn't register with them why anyone would do these things when we have all these machines that can do it for us, all these systems that promise to make life easier while slowly strangling us. But those machines are expensive to buy and expensive to maintain, and when they break (not if, when) you're left helpless until you can afford to replace them or pay someone else to fix them.

Physical vs. Digital Survival

Beyond CDs, we choose physical items over digital whenever possible, like choosing real food over synthetic substitutes. Books instead of e-readers that can be updated remotely to remove content that becomes inconvenient to the people who control the platforms. Maps instead of GPS that can be turned off or manipulated by people who want to control where you go. Tools instead of apps that stop working when the company goes out of business or decides your device is too old to support.

For me, I worked as a GIS professional making maps for a living: even for Supreme Court cases. Choosing physical maps for our personal use isn't ironic or contradictory. It's just using the best tool for the job, the tool that works when everything else fails. West Virginia doesn't have cell service in deep woods or some hollers, places where you need navigation most. You can download maps offline, but what if the battery doesn't work or you drop your phone in water or the software glitches or the satellites go down?

Physical maps don't require power, don't break when dropped, and don't stop working when technology fails or when someone else decides they should stop working. They're not "smart" in the way that modern devices are smart: they can't be updated remotely, they can't track your location, they can't serve you ads or collect data about your movements. They're just tools that do one job well, forever, without asking for anything in return.

The difference in how these things feel and function is real, like the difference between breathing air and breathing through a machine. A physical book doesn't run out of battery or stop working when the company that made your e-reader goes out of business. A paper map doesn't lose signal or lead you off a cliff because the algorithm decided you needed to see an advertisement. A CD doesn't disappear when a streaming service changes its catalog or when your internet goes down or when you can't afford the monthly subscription anymore.

The Acceleration Toward Helplessness

AI is accelerating the decline of self-reliance like pouring gasoline on a fire that's already consuming everything we need to survive independently. Every new tool promises to make us more capable while actually making us more dependent, more helpless, more vulnerable to the whims of systems we can't control. The solution is simple but radical: pull back from the edge of the cliff before you fall off. Take one thing and do it for yourself. And then another. That's how you reclaim ground that's been stolen from you one convenience at a time.

Convenience has made us dependent, like a drug that feels good while it slowly kills you. Or in my case, that extra-large glass of whiskey after a long day where ruining your liver becomes a nightly ritual. You don't have to participate in your own destruction. You don't have to be part of a world where ease means erasure, where convenience means incompetence, where every labor-saving device takes away another piece of your ability to survive without the system that's designed to consume you.

One step toward independence is enough to start. Wash your own dishes like you're reclaiming your own hands. Line-dry your laundry like you're remembering how to use the sun. Cook a meal from scratch like you're learning to feed yourself instead of being fed by strangers. Buy a CD like you're owning something real instead of renting everything digital. Process your own meat like you're taking responsibility for the life that sustains your life.

You'll feel it in your bones: the satisfaction of capability, the pride of self-sufficiency, the security of not needing other people's systems to meet your basic needs. Pull back from the conveniences that are slowly killing you. It's simple and free. Just pull back from one thing. Anything. Do it yourself. Do it in the way that offers real independence from a system designed to consume you whole. One bite at a time.

Chapter 4: The Numbers Don't Lie

The promise was mathematical. Follow the formula, get the result. But when we actually ran the numbers on the life we were supposed to want, the math didn't work.

We were carrying two mortgages with expenses totaling over $4,000 per month just for housing. Before utilities, before maintenance, before property taxes, before homeowner's insurance. Just the base mortgage payments. Then add everything else: car payments, insurance premiums, utilities, internet, cell phones, subscription services, groceries, gas, maintenance, repairs, unexpected expenses.

We had two six-figure incomes and were comfortable with the money we had left after each paycheck, but that's when the real question hit us: where does the spending end? We had accumulated all this stuff, so what's next? When do we become paycheck to paycheck despite making good money? What job will I need for that level of spending? How much more time am I going to lose because of it?

The education investment was supposed to pay off. I have a bachelor's and master's degree. Alexys continued through her doctorate. But when we looked at the actual return on investment, including opportunity cost, I feel like I would have been better off mentally and financially without my degrees. My GIS profession is dying due to technology and AI. Maps that used to require specialized training and expensive software can now be generated automatically. The skills I spent years developing are becoming obsolete. It is a hard pill to swallow to say that out loud. It's even harder to say to you, the reader. That career once defined me. It was a death I mourned for a long time.

Alexys's advanced education opened doors, but at what cost? Years of student loans, years of delayed income, years of stress and academic pressure. The debt service alone ate up much of the income advantage the degrees provided.

The retirement mythology was the biggest lie. That magic number, one million dollars in each of our accounts, was supposed to guarantee freedom. But when you do the actual math on what one million dollars gets you in retirement, especially with healthcare costs, it's barely survival. If you're lucky enough to get 4% annual returns without market crashes, that's $40,000 per year before taxes. Try living on that with medical expenses, prescription costs, and inflation eating away at purchasing power over twenty or thirty years.

We were supposed to believe that putting money into accounts managed by strangers, invested in companies we didn't understand, in markets we couldn't control, was somehow more secure than being able to feed ourselves.

The hidden costs of "doing everything right" were enormous. Commuting costs: gas, car maintenance, parking fees. Work clothes that you only wear to impress people you don't like. Eating out because you're too tired to cook after long days at jobs that drain your soul. Stress-related healthcare costs. The premium you pay for convenience because you don't have time to do things yourself.

Coffee and breakfast on the way to work instead of eating at home. Lunch out because the office doesn't have a kitchen. Dinner out because you're too exhausted to cook. Cleaning services because you don't have time to clean your own house. Lawn services because weekends are for recovering from the week. Each convenience seems small, but they add up to hundreds or thousands of dollars per month.

Then there were the costs of projecting success. The right car, the right neighborhood, the right clothes, the right vacations. All the money spent on looking like you were winning at a game that was rigged against you.

When we finally decided to liquidate everything and make the farm happen, we wiped over half a million dollars of debt off the books in one brutal, liberating move. Mostly mortgages: we didn't have much car debt or credit card debt because we were "responsible." But the mortgages alone were crushing us.

It was brutal because it represented a lifetime of work, all the stuff we had accumulated, all the equity we had built. The finality of it was terrifying in the moment we were signing the closing documents. But the liberation came later, once we could really feel what it meant to not owe anyone anything.

My mom got shitty with us about the decision. She couldn't understand walking away from "investment" properties and "building wealth." She said she'd never come visit us at our new property that is now our farm. That's the cost of breaking free: some people will never understand. My family no longer speaks to me. That is probably for the better.

The timeline was lightning fast once we committed. We started looking for land as a retreat in August 2023. By October, it shifted from weekend getaway to "we don't want these houses anymore." We closed on our farm property in December 2023 and sold everything else by February 2024. Four months from decision to complete exit.

We priced everything aggressively to get out fast. We even took a hit on our dream home that we had purchased just one year prior and had to pay money to get out of it. Tens of thousands of dollars in concessions we gave up to get out quickly and priced it the same as we bought it for. That's how serious we were about checking out of this system.

We had to get creative with financing. We sold lots of things and cashed out accounts, including my retirement account. We took the penalties and taxes because freedom was worth more than the theoretical future value of money we might never see.

The real math of our new life is simple: we went from over $4,000 per month just in mortgage payments to zero. Our monthly expenses dropped dramatically. We don't need six-figure incomes to survive anymore. We don't need to work jobs that slowly kill us to afford a lifestyle that was making us miserable.

The numbers don't lie. The traditional path is mathematically broken for most people. The costs keep rising while the benefits keep shrinking. The promises are based on assumptions that no longer hold true. The math works for the system, not for the people trapped in it.

We did the math on freedom, and it added up.

Part II: Building the Exit

Chapter 5: Start Where You Stand

The decision didn't happen overnight, but once it started, everything moved fast like an avalanche that starts with a single pebble and ends by burying everything in its path.

When we started looking for land in August 2023, we were initially thinking about it as a retreat, a place to escape to on weekends when the weight of our "successful" life became too crushing to bear. Somewhere to breathe when the beautiful prison we'd built around ourselves felt like it was running out of oxygen. By fall we didn't want to go back. The retreat became an exit strategy, the weekend escape became a permanent evacuation plan.

That October, we had the conversation that changed everything, sitting in our "dream home" that felt more like a mausoleum filled with expensive furniture we'd bought to impress people whose opinions didn't matter. We were looking around at all our stuff like archaeologists examining the artifacts of a dead civilization, and we both knew we were done. We didn't want to maintain it like curators of our own suffering. We didn't want to pay for it like making payments on our own graves. We didn't want to live surrounded by it like prisoners decorating our cells to make captivity more bearable.

The Liquidation Death March

The liquidation had to happen fast like amputating a gangrenous limb before the infection killed the whole body. We priced both properties aggressively, willing to take hits that would have made our former selves sick with grief. We sold our dream home for exactly what we'd paid for it one year earlier, eating tens of thousands in closing costs, real estate fees, and concessions just to escape.

The market was okay, but we were desperate to be free, desperate enough to pay almost any price for the privilege of not owing our souls to banks that saw us as nothing more than payment streams with pulse rates. We weren't trying to maximize profit. We were trying to minimize the time between decision and freedom, between recognizing we were dying and actually starting to live.

Most people would see what we did as financial suicide, but we saw it as the only way to survive. Most people would look at what we were giving up: the houses, the equity, the appearance of success, while we looked at what we were gaining: the chance to live like human beings instead of dying like human resources.

The Financing Nightmare

We were trying to buy property while selling two others, coordinating closings, and managing cash flow when everything was tied up in assets we were liquidating as fast as possible. We had to get creative like criminals planning a heist, except we were stealing our own lives back from a system that had stolen them legally.

We sold personal belongings. We cashed out retirement accounts, took penalties and paid taxes we could have avoided if we'd been more patient. But patience would have meant staying trapped longer, and every day in the system was another day of our lives fed into a machine that gave back nothing but the promise of more suffering disguised as security.

The weight of desperation made every decision feel life-or-death, because it was. This wasn't financial planning. It was escape planning. This wasn't investment strategy. It was survival strategy. We weren't optimizing our portfolio. We were burning it down before it burned us.

Living in Limbo

We were living in a camper on our new property for those months between December and February, off-grid for part of it because we demolished two buildings and had to get utilities reconnected in the dead of winter. Us, our three dogs, and a cat, living off a generator in a West Virginia winter while trying to coordinate the sale of everything we'd spent years accumulating.

It was like living in purgatory, suspended between the life we'd left and the life we were trying to build, not knowing if we'd made the biggest mistake of our lives or the smartest decision we'd ever make. Every morning we woke up in that camper was another day of choosing uncertainty over certainty, choosing the possibility of failure over the guarantee of slow death.

The animals were our anchor to sanity, the only constant in a world where everything else was changing too fast to process. They didn't care about our financial decisions or our real estate transactions or our radical lifestyle changes. They just needed to be fed and loved and cared for, which gave us something solid to hold onto when everything else felt like it was dissolving.

February Freedom

By February 2024, we were free like prisoners walking out of jail after serving time for crimes we didn't commit. No mortgages. No car payments. No debt except what we chose to take on for the farm. We had wiped over half a million dollars of mortgage debt off the books and traded it for one acre of land and the chance to build something real.

The timeline was October decision to February freedom: four months to completely exit a system we'd spent decades building our lives around. It can be done fast if you're willing to take the financial hit and prioritize freedom over maximizing profit, if you're willing to lose money to save your life.

Most people aren't willing to take the loss, aren't willing to accept that escaping the system costs money because the system is designed to make escape expensive. They want to time the market, maximize their equity, wait for the perfect moment that never comes because perfect moments don't exist in imperfect systems.

The perfect moment is a lie told to keep you waiting. The system is designed to keep you waiting, always one more promotion, one more payment, one more year until you can afford to be free. But you can't afford to be free in a system designed to keep you enslaved, and waiting for permission to escape from your captors is just another form of captivity.

The Price of Freedom

We decided that freedom was worth more than money, that four months of intense stress and financial loss was better than decades of slow suffocation dressed up as success. That starting over with nothing was better than continuing to accumulate things we didn't want to support a life we didn't believe in.

Start where you stand, which for us was standing in quicksand that was pulling us under one inch at a time. You don't need permission from the system to leave the system. You don't need the perfect plan or the ideal circumstances or the approval of people who are still trapped in what you're trying to escape.

You need to decide that your life is worth more than your stuff, that your time is worth more than your possessions, that your freedom is worth more than other people's opinions about your choices.

The hardest part isn't the logistics of escape. It's accepting that you're going to take a financial hit, that you're going to lose money on the transaction, that you're going to walk away from equity and investments and opportunities that other people would kill for. People are going to think you're making a mistake, and maybe you are, but it's your mistake to make.

Better to fail at something you believe in than succeed at something that's killing you. Better to risk poverty with purpose than accept wealth without meaning. Better to choose your own destruction than let someone else destroy you slowly and profitably.

The system only has power over you as long as you believe you need its permission to live your life. Start where you stand, even if where you stand is in the middle of a burning building. Especially if where you stand is in the middle of a burning building.

The exit signs are there. You just have to be desperate enough to use them.

Chapter 6: Finding Your Land

Finding land when you've spent twelve years in federal land acquisition is like a recovering alcoholic knowing exactly which bars serve the best drinks. The knowledge is useful, but the context has changed completely. During my tenure with the federal government, I worked on projects that expanded some of the largest national parks in the country. Tens of thousands of those acres came through my hands and are now part of the millions managed and protected by the National Park Service. My work also included mapping billion-dollar broadband infrastructure projects. I spent years learning to read land like a second language. Its resources, infrastructure, rights-of-way, encumbrances, easements, boundaries, and potential were all part of the vocabulary. Now we were putting that knowledge to work for ourselves instead of for bureaucrats who couldn't find their own asses with GPS coordinates and a survey crew.

We had no idea what we wanted, which turned out to be an advantage because we didn't waste time chasing fantasies that don't exist in the real world. We wanted it to be pretty, not on top of neighbors who would call the county about everything we did. We preferred lots of mature trees, because planting trees and waiting thirty years for shade seemed like planning for a future we might not live to see. It needed to be connected to utilities so we could reestablish them without having to run power lines through pristine wilderness like pioneers with credit cards.

We had exactly two deal-breakers: internet access so we could run our business, and proximity to a hospital so Alexys could get her monthly MS infusion without turning medical appointments into weekend camping trips. Everything else was negotiable, which meant we weren't picky enough to eliminate good properties for stupid reasons like paint colors or kitchen layouts in houses we planned to demolish anyway.

We were willing to look anywhere in West Virginia, which probably seems random to people who think geography is about convenience rather than freedom. We aren't from West Virginia, but we love it more than any place on earth because its beauty is unmatched by anywhere we've traveled, and its people generally leave you alone to live your life instead of trying to manage your choices for you. Mountains don't care about your politics, and trees don't report you to homeowners associations for having chickens in your backyard.

The Search That Wasn't Really a Search

We searched on Zillow like everyone else, but I was filtering properties through decades of professional experience in evaluating land for purposes that actually mattered instead of aesthetic preferences that serve no practical function. We weren't looking for move-in ready homes. We didn't care about granite countertops or open floor plans designed by people who think kitchens should look like showrooms rather than places where food gets prepared. We were looking for potential that didn't depend on someone else's vision of what rural life should look like.

We found our property through Zillow and contacted the listing agent immediately. We didn't want to play games with multiple agents or try to negotiate around commission structures that serve nobody except real estate professionals. Speed and efficiency mattered more than saving a few thousand dollars on agent fees. We also wanted the listing agent to get the full commission because cooperative agents work harder when they're getting paid properly for their time and expertise.

Our first impression was that the house was a dump that smelled like mold and had water damage throughout, but we were planning to demolish it anyway so the condition of the structure was irrelevant as long as the utilities were connected. We didn't care that the foundation was sinking into a swamp. The property itself was an absolute dream once we looked past the junky home and garage that represented someone else's idea of rural living, someone who had clearly given up on maintenance and improvement years before putting it on the market.

We fell in love with the land after we ignored the buildings and started evaluating the property for what it could become rather than what it currently was. I used my GIS background to analyze the terrain like a professional instead of walking around pointing at pretty views like tourists. I pulled up USGS Web Soil Survey data to determine soil composition and drainage patterns, topographic maps to understand how water would shed during heavy rains, and decades of historical aerial imagery to see how the land had evolved over time and what previous owners had done to improve or damage it.

Nobody knows this land better than us now, not even the folks who lived here for sixty years before we bought it. They lived on it, but they didn't study it like professionals whose livelihood depends on understanding every aspect of terrain, drainage, soil quality, and environmental factors that affect land use and development potential. We knew more about our property before we closed than most people learn about their land after living on it for decades.

Due Diligence for People Who Aren't Idiots

We waived all inspections and bought the property as-is because we had enough professional experience to evaluate the important factors ourselves. We didn't need to pay inspectors to tell us things we already knew about properties that we planned to modify extensively anyway. Inspections make sense for people buying finished homes where hidden problems could cost tens of thousands to repair, but they're mostly expensive paperwork exercises for people buying fixer-uppers that they plan to gut and rebuild from scratch.

What concerns did we have? Too many to count, but we dove in headfirst because concerns are problems that can be solved, not permanent obstacles that prevent action. Every piece of rural property has problems: access issues, utility limitations, drainage challenges, neighbor conflicts, zoning restrictions, maintenance requirements that would horrify suburban homeowners who think property ownership means calling management companies when something breaks.

Gas, water, and electric were already connected, which saved us months of dealing with utility companies that treat new rural customers like inconveniences rather than revenue sources. We permanently removed the gas lines and meter from the property because gas is another utility dependency that can be controlled by people who don't care about your survival or comfort, and we didn't want explosive utilities that could kill us if they malfunctioned or got sabotaged by people who profit from our continued dependence on their services.

Surprises happened regularly, but most were entertaining rather than expensive. We keep finding marbles in the dirt everywhere we dig: plant a tree, find marbles; hoe the garden, find marbles; dig post holes, find marbles. Turns out the original owner worked at a marble factory and used them in his slingshot to hunt squirrels and other small animals around the property. Decades later, we're still discovering his ammunition scattered throughout the soil like archaeological evidence of someone who took small game hunting seriously and had access to unlimited projectiles.

The Purchase That Actually Worked

We didn't finance anything because mortgages are just another form of slavery with paperwork that makes it look legitimate. We bought it with cash: $60,000 for one acre with a house and two garages. Which seems impossibly cheap to people who think real estate prices in overpriced metropolitan areas represent actual value rather than artificial scarcity created by zoning laws and development restrictions that serve existing property owners at the expense of everyone trying to buy homes.

No mortgage means no monthly payments to banks that profit from your labor, no interest charges that double the actual cost of the property over thirty years, no risk of foreclosure when economic circumstances change, no debt service that requires you to maintain employment with people who treat you like livestock that generates revenue. Cash purchases mean you own the property immediately and completely, not gradually over decades while paying rent to financial institutions that call it ownership but retain the right to take it back if you miss payments.

No complications occurred because cash deals eliminate most of the bureaucratic complications that make financing a nightmare of paperwork. Underwriting, appraisals, and approval processes are designed to extract fees while creating opportunities for deals to fall apart when banks change their lending criteria or discover new ways to profit from other people's desperation. We closed in two weeks from offer to ownership, even during the holidays when most people and businesses are too distracted by seasonal obligations to focus on completing transactions efficiently.

The closing process was straightforward because we weren't dealing with mortgage companies, title insurance requirements, or inspection contingencies that create multiple opportunities for problems and delays. We made an offer, it was accepted, we verified ownership and title status, we brought a check, we signed papers, we got keys. Simple transactions for people who aren't trying to leverage other people's money to buy things they can't afford with income they might not have next year.

The Agent Who Actually Worked

We contacted the listing agent directly because she had already done the work of marketing the property and showing it to potential buyers. We wanted her to receive full compensation for her efforts rather than splitting commission with buyer's agents who would add complexity and potentially delay a transaction that we wanted to complete as quickly and simply as possible.

Real estate agents work harder and more cooperatively when they're getting paid properly for their time and expertise. This is especially true in rural markets where properties don't sell quickly, and agents often work for months to complete transactions that generate less commission than suburban home sales that close in weeks. We wanted efficient service from someone who had incentive to make the deal happen rather than someone who might prefer to show us more expensive properties that would generate higher fees.

The agent was competent and professional, which seems rare in an industry where most practitioners treat buyers like walking commission checks rather than people making major life decisions that affect their financial security and future wellbeing. She answered questions honestly, provided relevant information about the property and area, and facilitated the closing process without creating artificial drama or urgency designed to pressure us into making decisions that served her interests more than ours.

Rural real estate is different from suburban markets where agents show houses that look identical and buyers choose based on minor differences in fixtures and finishes. Rural properties are unique, and buyers need agents who understand land use, utility access, zoning restrictions, and practical considerations that affect rural living. Buyers in rural markets ought to avoid agents who specialize in staging and curb appeal for people who think property value is determined by aesthetic preferences rather than functional utility.

What We'd Do Differently, Which Is Nothing

We wouldn't do anything differently because we approached the purchase strategically rather than emotionally. We used professional expertise to evaluate the property thoroughly, and structured the transaction to serve our interests rather than conforming to standard practices that serve other people's profit margins. We bought exactly what we needed for exactly what we wanted to pay, closed quickly and efficiently, and took ownership of property that has provided everything we expected and more.

Most people make real estate purchases based on fantasy rather than reality. They buy properties that look good in photos but don't serve their actual needs, in locations that seem desirable but don't support their lifestyle goals, with financing that traps them in employment situations that prevent them from enjoying the properties they're paying for. They buy dreams instead of land, and they get nightmares instead of freedom.

We bought dirt, potential, utility connections, and legal ownership of space where we could build the life we wanted. Build it without asking permission from neighbors, HOAs, municipal authorities, or anyone else who profits from controlling other people's choices about how to use property they ostensibly own. We bought the right to make mistakes, build ugly structures, raise animals, grow food, and live according to our values rather than other people's aesthetic preferences or social expectations.

The only thing we might have done differently was buy more land, but that would have required more money or financing. We prioritized debt elimination over acreage maximization because debt is slavery, regardless of how much land it helps you control. That control is temporary. The lenders always come back for their money, plus interest, plus penalties, all for the privilege of using their capital to buy something you couldn't afford with money you never actually had.

Finding land isn't about finding the perfect property. It's about finding good enough land that you can shape to fit your needs and values without interference from people who think they know how you should live. Perfect doesn't exist, good enough is available everywhere. Improvement is possible anywhere if you're willing to work for it. Not by paying others to do the work, but by picking up the tools yourself. With time, effort, and the willingness to learn the skills most people were taught to believe they couldn't learn.

The search ended when we found land that met our minimum requirements and offered potential for development according to our long-term goals. Everything else was details that could be solved with money, labor, and time. All resources that we had available because we weren't wasting them anymore on mortgage payments, car loans, credit card debt, and other financial obligations that serve other people's interests rather than our own survival and independence.

Chapter 7: The Demo and Setting Up Infrastructure

The demolition of the house and one of the garages started while we were hours away, packing up the remnants of our suburban existence like refugees preparing to flee a war zone that most people call normal life. It's important to know that not every analogy is literal, but we were in fact running for our lives. Before we ever earned a single dollar, we lived through substantial trauma. Physical. Emotional. Relentless. And we worked our asses off to claw our way out of it. We broke through, yes, into six-figure salaries, but only to find ourselves in a more expensive version of the same shit hole. Just with better amenities. So understand this: we weren't escaping comfort. We were escaping collapse.

We knew the demo had begun when our soon-to-be neighbors started texting us about the giant track hoe that appeared on our property like an alien invasion, followed by the simple question: "Are you demoing the house?" Which was their polite yet nosy way of asking whether we were the new crazy people who bought the moldy dump just to tear it down, or whether some contractor had made a very expensive mistake about which property to destroy.

Two days later, we had a blank slate. What was once a house was now a straw-covered bowl in front of the little blue garage we'd soon call our home. The concrete block garage that scarred the beautiful neighborhood was now a flat area ready to be reclaimed by nature. A camper to live in and a whole lot of freedom, we arrived a week later to a property ready for us to rebuild.

The garage-to-home conversion took three months. During the conversion we lived in a camper through a West Virginia winter like pioneers with RVs, and unrealistic expectations about the romance of off-grid living.

The work was extensive and mostly happened in winter weather that made every task take twice as long and twice as difficult. I'll never forget suffering through freezing temperatures when your hands quit working. When you have to shovel snow before you spend the entire day in a cold damp garage wondering if you completely fucked up.

It was a shitty experience, to say the least, but it was temporary discomfort in service of permanent improvement rather than permanent discomfort in service of other people's profit margins. This is what most people accept as normal employment and housing arrangements. Three months of voluntary hardship beats thirty years of involuntary servitude, even when the hardship includes hauling water, emptying waste tanks, and running generators for electricity like camping without the option to go home when the novelty wears off.

The Gas Line Removal: Safety First, Dependence Never

Prior to the demo, we removed the gas line for safety and never intended on putting it back in. Gas leaks kill people and explosions destroy homes. But more than that, gas is just another utility dependency. Another system controlled by people who don't care whether you live or die, as long as you pay your bill and stay quiet.

Gas companies can shut off your service when you can't pay, when your usage looks suspicious, when they need to do maintenance, or when it becomes profitable to restrict supply and blame it on external factors. Price hikes follow, and their quarterly earnings go up. Gas appliances rely on gas delivery. Gas delivery depends on infrastructure controlled by people who care more about shareholder returns than whether your house stays warm.

We chose to eliminate gas dependency entirely rather than maintain it as a backup system because backup systems require maintenance and monitoring and create additional points of failure that can compromise your independence when you need it most. Every utility connection is a potential control mechanism that other people can use to influence your behavior or extract money from your basic survival needs.

Removing gas lines permanently also eliminates ongoing safety monitoring and maintenance requirements for equipment that can kill you. It should be well known that damaging or tampering with gas lines creates immediate life-threatening emergencies. These emergencies force property owners to pay emergency repair costs while living without heat or hot water until the problems get fixed by technicians who charge premium rates for urgent service calls.

Electrical Challenges: When Professionals Get Sick at the Worst Times

We also had to remove the electric service before the demo because it was connected to the house. Getting electric service reestablished on the garage (that is now our home) was a challenge because utility companies treat reconnections like new installations that require inspections, permits, updated wiring. It also requires compliance with current codes that didn't exist when the original electrical service was installed decades ago by people who prioritized functionality over regulatory compliance.

Living in the camper in the winter, we desperately needed electricity so we could plug in our temporary housing and run basic systems for survival and comfort during the renovation process. The electrician got sick and the weather was freezing, which meant delays and rescheduling and waiting for conditions that allowed safe electrical work in outdoor conditions. Conditions that could kill workers who made mistakes while installing systems that could kill residents if they were installed incorrectly.

The delay was frustrating but not catastrophic because we planned ahead. We had backup systems and redundant plans for heating, lighting, and power generation that didn't depend on any single source or service provider. Generator power worked for essential needs while we waited for grid connection, and the camper's systems provided basic services that kept us alive and functional while permanent infrastructure got installed.

Professional electrical work costs more and takes longer than doing it yourself. But electrical mistakes can burn down buildings or kill people. That's why we hired licensed electricians instead of gambling with work we weren't qualified to do and liability we couldn't afford. A fire. An injury. One wrong wire could cost more than any installation fee. And honestly, what's the point of having this amazing new property if we're fucking dead? We don't take risks like that. Not now. Not ever.

Waste Management: The Unglamorous Reality of Rural Living

The sewer was disconnected for the demo. Waste disposal meant driving to a nearby RV dump station a few times a week. Most people are used to suburban sewage systems that handle everything automatically and invisibly. They never have to think about where their shit goes, who deals with it, or what happens when those systems fail during a storm or an infrastructure collapse.

RV dump stations are exactly as pleasant as they sound. Not pleasant at all. But still better than overflowing tanks or illegal dumping that creates health and environmental hazards for people trying to live responsibly while transitioning from suburban dependency to rural self-sufficiency. Emptying waste like that is a humbling reminder that every modern convenience depends on infrastructure most people never see or think about until it stops working.

The drive to the dump station became part of our routine, something we had to plan around tank capacity and weather conditions. Unlike suburban systems that run no matter the weather, dump stations are often closed when conditions get bad. That's exactly when driving becomes more dangerous with a vehicle carrying shifting liquid weight that changes how it handles on the road.

Each dump trip was a reminder that waste is directly connected to consumption. Use less, and there's less to haul. That becomes obvious when you're personally responsible for transporting and dumping everything you produce. Most people flush into systems that make it all disappear, like infrastructure is performing some kind of magic. But that infrastructure costs money, breaks down, and creates dependency on people who profit from handling things you could deal with yourself if you were willing to confront the basic realities of human life.

Internet Access: The Rural Reality of Digital Infrastructure

There weren't many good internet providers in our area, but we needed something stable and reliable. We were both working full-time during the transition, before I quit my job to focus on the farm. Rural internet options are limited because telecom companies prioritize suburban markets that are cheaper to serve and more profitable. Rural areas require more infrastructure to reach fewer people, which means less revenue per mile. So they don't bother.

We researched the available options and chose the most reliable service we could afford. It wasn't the fastest or the cheapest, but it was the most consistent for business use. We needed internet that could handle video calls, file uploads, and constant connectivity without being interrupted by weather or equipment failures. The cheaper services were designed for residential customers who use the internet for entertainment, not for income generation.

Rural internet costs more and works less reliably than suburban internet, but it's sufficient for business needs if you adjust expectations and develop backup plans for connectivity failures. These failures happen more frequently in rural areas where infrastructure is older and maintenance is deferred because profit margins are lower for companies that serve dispersed customer bases rather than dense suburban markets.

The Garage Conversion: Making Something From Nothing

The original two-car garage looked like a two-car garage. Obvious, but worth stating. People ask detailed questions about things that are exactly what they sound like, as if simple descriptions must hide some deeper complexity. As if they've never seen the basic structures found across rural America. Garages built to store vehicles and equipment. Nothing more.

The garage is 740 square feet with three garage doors and one pedestrian door, which seemed like a lot of space until we started planning how to convert it into living space that would meet our needs. It needed to meet the needs for sleeping, cooking, working, and storing all the possessions we were keeping from our previous life. Converting garage space to living space requires insulation, heating, plumbing, interior walls, flooring, lighting, and ventilation systems that weren't needed when the space was used for vehicle storage.

When we first closed on the property, we drove to take possession after signing papers and pulled one of our campers into the garage to spend our first night on the property. We were camping inside the garage like people who couldn't decide whether they wanted to be indoors or outdoors and chose both simultaneously. It was an amazing but scary feeling to own property that we could use however we wanted without asking permission from landlords, HOAs, or municipal authorities who normally control how people use space they think they own but actually rent from various regulatory bodies.

The conversion process required professional help for electrical, but we did most of the other work ourselves. Labor costs more than materials and we had more time than money during the transition period when we were liquidating assets from our previous life while building infrastructure for our new life.

All the interior work, from the guts to the finishing, takes time but isn't technically difficult if you're willing to learn basic construction skills. Skills most homeowners used to have before they started hiring contractors for everything. Previous generations did this work themselves with hand tools, common sense, and trial and error. They learned by doing, not by paying for someone else to teach them or do it for them.

Infrastructure as Independence

The entire garage became our home. We didn't need separate rooms for every activity the way suburban houses divide space into single-use areas. That kind of layout wastes square footage that could be used more efficiently by people who care more about function than appearances. A lot of interior design exists to impress guests who judge homeowners by décor instead of competence or satisfaction. That doesn't mean our home is a shithole. It's small, it's adorable, and we love it. You probably would too.

Open floor plans work better for people who actually live in their space instead of entertaining people they don't like in rooms they don't use, filled with furniture they bought to impress people whose opinions don't matter. None of that improves their wellbeing, happiness, or security. We needed space to sleep, cook, work, and store what we use. So we designed the conversion to maximize utility and keep costs and maintenance low.

The conversion was our first major project in building infrastructure that served our needs instead of following standard expectations. Most of those expectations are unrealistic anyway, shaped by people who sell house plans, building materials, and renovation services. They push the idea that homes should look like magazine photos instead of function like tools that support the way people actually live.

Every part of the infrastructure development was a learning experience. It built skills, built confidence, and cut our dependence on contractors and professional services. Those services cost money, require scheduling, and cause delays when other people's priorities don't match your timelines. Each finished project was proof that we could solve problems and build solutions without relying on anyone else's competence. We also stopped depending on whether someone was available or willing to work for what we could afford.

Infrastructure development is an ongoing process. It keeps us busy and grounded in practical work that produces visible results and real improvements to how we live and operate. Building and maintaining our own systems is more satisfying than paying others to manage infrastructure we rely on but don't understand and can't fix when it breaks at the worst possible time.

What we built serves our needs and values. It doesn't follow standard practices designed to protect profit margins or appeal to shallow aesthetic trends. Every system was built for reliability, simplicity, and ease of maintenance. We didn't waste money on features that look impressive but don't improve function, reduce costs, or increase independence from outside systems. And yes, we used to have money and expensive taste, so it's still pretty damn cute.

Chapter 8: The Camper Months

The first night in that camper was terrifying.

We arrived at the property in late January, in the afternoon when the winter light was already fading. It was cold: the kind of West Virginia cold that cuts through whatever you're wearing and reminds you that nature doesn't care about your comfort or your plans. We had mere hours to figure out where we could safely park the camper, move our necessities into it, power up the generator to get the heat going, and get our dogs and cat comfortable. Every minute that passed more anxiety creeped in. Could we pull this off?

Dinner was a couple of soup cans cooked in the microwave with bread to wipe it up. Not exactly the celebratory meal you'd expect for your first night of freedom, but it was what we had, and it was ours. No restaurant bills, no delivery fees. Just simple sustenance in our new reality.

The sleeping arrangements were an exercise in making the best of a tight situation. Goose the cat, Clemmy our Lab, and Latty our little dog all slept in the bed with us. Moira, our Bernese Mountain Dog, slept on the floor because it was nice and cool; you can imagine what a camper floor feels like in January. But she seemed content, and we were all together, which was what mattered.

The 22-Foot Reality

The camper is small. A 22-foot travel trailer including the tongue, weighing about 4,000 pounds. Most RVers will know it's not very big. A Murphy bed that folded down from the wall. A small kitchen with basic appliances. A bathroom with a foot-flush toilet, shower, and small sink. That was our entire living space for three months.

We navigated such a tight space with love and compassion for one another. We were fully in it together: had to be. There was no privacy, no personal space, no room to storm off when tensions got high. Every movement had to be coordinated. Every decision affected both of us immediately.

When one person was cooking, the other had to stay out of the kitchen area. When someone needed to use the bathroom, the other person had to plan around it. When the bed was down, the living area disappeared. When we needed to access storage, everything had to be moved around carefully in the limited space.

But instead of driving us apart, the forced proximity brought us closer. We learned to communicate more directly, to be more considerate of each other's needs, to find humor in situations that could have been maddening. We developed systems and routines that maximized the small space and minimized conflict.

The Walmart Water Runs

The water situation was brutal and humbling. Going to Walmart to fill up our 7-gallon water cans was not only challenging but drew attention from other shoppers. We went from luxury to filling up our shower water at a retail store in the back where the water dispenser was located.

It was horrible. Navigating a Walmart with multiple containers, finding the water station, waiting for other customers to finish, filling up while people stared at you with a mixture of curiosity and pity. The whole process took close to a couple of hours, and we had to do it a couple of times per week.

People would watch us loading case after case of water into our cart, clearly wondering what we were doing with so much water. Some would ask if we were having a party or stocking up for an emergency. We'd give vague answers about living in a rural area with water issues, which was technically true but didn't capture the full reality of our situation.

The drive back to the property with a vehicle full of water was always anxious. What if we got a flat tire? What if the containers leaked? What if we spilled everything and had to start over? The weight of the water, figuratively, not literally, changed how I drove: carefully, knowing that our supply of clean water for the next few days was entirely dependent on making it home safely and me not fucking it up.

Unloading and transferring the water to the camper's storage tanks was tedious work. Heavy lifting, careful pouring, making sure nothing spilled because every drop was precious.

It was a constant reminder of how much we'd taken running water for granted. Turn a handle, water appears. No thought about where it came from, how it got there, or what would happen if it stopped flowing. Now every gallon was earned through effort and planning.

Generator Dependence

We had never used a generator before that very moment we fired it up for the first time, and suddenly our lives depended on it working. The anxiety of starting it up for the first time was overwhelming. What if it doesn't start? What if it runs out of gas in the middle of the night? What if we can't figure out how to keep it running?

The generator worked fine, but the psychological pressure was intense. This wasn't camping for fun where you could pack up and go home if things went wrong. This was our life now. We relied on that generator to live, with no backup plan other than trying to find a last-minute Airbnb that would accept animals, which, in rural West Virginia in winter, was not a reliable option.

We had to go buy gasoline for the generator just to stay alive in our camper. Regular trips to gas stations, carrying fuel cans, calculating how much we needed to get through cold nights and cloudy days when solar wasn't an option. The sound of the generator became the soundtrack of our early days: a constant mechanical hum that meant warmth, light, and the ability to cook food and charge devices.

But it was also a reminder of our vulnerability. Everything depended on that one machine working properly. If it failed completely, we'd be in serious trouble fast. No heat in freezing temperatures. No way to cook food. No way to charge phones for emergencies. No water pump to move water from storage to faucets.

The Animals Came First

Lucky for us, our animals are amazing. We made sure to prioritize them by immediately fencing in an area where they could run and play. We put our cat Goose in a backpack to come on walks with us.

They are our everything, and we weren't going to let our transition disrupt their lives more than necessary.

The animals adapted better than we did in some ways. Dogs are remarkably resilient, and as long as they're with their people, they can handle almost anything. The cat was less thrilled with the confined space, but even Goose seemed to understand that this was temporary and that we were all in it together.

We took them on drives regularly to break up the monotony and give everyone a change of scenery. Long walks in the woods, even in cold weather, became daily necessities for both human and animal sanity. Fresh air, exercise, and space to roam made the return to the cramped camper more bearable.

The fenced yard was one of our first priorities, even before getting utilities reconnected. The animals needed space to be animals, and we needed to know they were safe and contained and had an area to burn off steam. They desperately needed it while we worked on infrastructure and renovation projects that kept them confined to the camper or sections of the garage we weren't renovating.

Learning What We Could Handle

The camper months taught us what we actually needed versus what we thought we needed. We didn't need much. Water, heat, food, and each other. Everything else was luxury or distraction.

Critical thinking and troubleshooting became essential skills. We'd used those abilities in our corporate jobs, but never in this type of way. Never with our immediate survival depending on the outcome.

When something broke or didn't work, we couldn't call a service or wait for someone else to fix it. We had to figure it out ourselves, right away.

Three months of that life taught us that we could handle much more than we thought we could, and need much less than we'd been told we needed. By the time we moved into our renovated home, we appreciated everything differently. Running water felt like magic. Reliable electricity was a gift. Having enough space to move around was luxury beyond description.

But we also learned not to take any of it for granted. We kept the skills we'd developed and the systems we'd created. We still monitor our water usage carefully. We still maintain backup systems for heat and power.

The camper months weren't something we endured: they were something we conquered. They proved that we could survive on our own terms, even when those terms were harsh and unforgiving. They gave us confidence that whatever came next, we could handle it.

Most people would have quit. Most people would have crawled back to their mortgages and their jobs and their comfortable dependence. We stuck it out because the alternative, which was going back, was worse than any temporary hardship.

The camper months ended, but the lessons stayed. We learned that comfort is earned, not owed. That resilience is built, not born. That freedom costs more than money, but it's worth whatever you pay for it.

Chapter 9: Own What Matters

Tool philosophy became simple: don't rent anything because if you need it, it might not be there. Rental companies close. Equipment breaks down. Delivery trucks get stuck. Weather prevents pickup. When you need a tool for something essential, you need to own it or have an alternative that doesn't depend on other people's schedules or availability.

We bought tools that could handle multiple jobs. A good shovel that could move dirt, snow, or gravel. A reliable chainsaw that could cut firewood, clear fallen trees, or process lumber. Hand tools that didn't require electricity or fuel. Basic tools that had been solving problems for hundreds of years before power tools existed.

But owning tools is different from owning stuff. Tools solve problems. Stuff creates problems. A hammer builds things. A decorative vase collects dust. A cast iron pan cooks food for decades. A fancy coffee maker breaks down and becomes trash.

We learned to distinguish between assets and liabilities. Assets make your life better or easier or more independent. Liabilities require maintenance, storage, insurance, or ongoing costs. Most of what we'd owned in our previous life were liabilities disguised as assets.

The income property we thought was building wealth was actually a source of constant stress and expense. Tenants, maintenance, insurance, property management, vacancy periods. It wasn't making us rich; it was making us busy and anxious.

The second car we thought we needed for convenience was mostly a parking problem and an insurance payment. We barely drove it, but we paid for it every month anyway.

The campers we bought for recreation spent most of their time depreciating in storage while we paid storage fees.

Now we own things that directly support our ability to live independently. Animals that produce food. Tools that help us maintain our infrastructure. Equipment that helps us process what we grow and raise. Land that provides space, resources, and peace.

Everything else is secondary. We don't buy things unless they solve a specific problem we're actually having. We don't upgrade unless something is actually broken. We don't accumulate for the sake of accumulating.

Storage space is limited, which helps with decision making. If something doesn't earn its place in our limited space, it doesn't belong here. Every item has to justify its existence by being useful, beautiful, or meaningful, and ideally all three.

We've become ruthless about eliminating redundancy.

Maintenance is part of ownership. We fix things instead of replacing them. We clean tools after using them. We store things properly to prevent damage. We learn how systems work so we can maintain them ourselves instead of hiring services.

The goal isn't minimalism for its own sake. The goal is intentionality. Every thing we own should serve a purpose that aligns with our values and supports our independence. If it doesn't, it's just weight we're carrying.

Quality matters more than quantity. One excellent tool that lasts decades is better than multiple cheap tools that break and need replacement. One warm coat that handles all weather is better than multiple specialized jackets that each handle specific conditions.

We don't buy things to impress people or signal status. We buy things that work. Things that last. Things that help us do what we want to do without depending on other people's systems and schedules.

This isn't about being cheap or depriving ourselves. It's about being selective and intentional. It's about owning things that support our independence rather than undermining it.

The hardest part is unlearning the consumer mindset. We were trained to buy solutions to problems we didn't have, to upgrade things that were working fine, to accumulate things that made us feel successful rather than actually making us more capable.

Now we ask different questions before buying anything: Does this solve a real problem I'm currently having? Can I fix or maintain this myself? Will this make me more or less dependent on other people's systems? Is this the simplest solution that will work?

Most of the time, the answer is no. Most of the time, we already have what we need or can make do with what we have. Most of the time, buying something new creates more problems than it solves.

Own what matters. Everything else is just noise.

Chapter 10: Cut Every Cord You Can

The subscriptions were the easiest place to start cutting, like removing leeches that had been feeding on us so long we'd forgotten they were there. We had expensive cell phone plans with unlimited data that we barely used, paying for bandwidth we didn't need to access information designed to make us anxious, broke, and distracted from what actually mattered.

Now we pay $30 a month for two lines. It feels less like buying into surveillance and more like paying for the ability to communicate. We don't buy phones every year on upgrade plans like upgrading prison cells, and we take care of what we have because what we have works for what we need. A phone is a tool, not a status symbol, or an entertainment device, or a tracking system that reports your location and habits to people who profit from your personal information.

We don't stream music anymore. We buy CDs because we want to own the music, support the artists directly instead of feeding platforms that pay musicians almost nothing while extracting maximum profit from their creativity. Don't make the mistake of depending on platforms that can disappear, change their catalogs, or raise their prices whenever they want more money from people who just want to hear songs they like.

We don't have gym memberships because we get our exercise from farm work like human beings instead of hamsters running on wheels in buildings designed to make fitness feel like punishment. We don't pay for software we don't need, software that stops working when you stop paying, software that updates itself to become worse while demanding more money for the privilege of being disappointed by products that used to work.

We don't use DoorDash or any delivery services because cooking and eating are basic human activities that we're capable of managing ourselves without paying markup to middlemen who profit from our laziness and exhaustion.

The Death by a Thousand Small Cuts

The subscription economy is designed to make you forget you're paying for things, like being slowly poisoned with doses too small to notice until you're already dead. Small monthly charges that seem reasonable individually but add up to hundreds of dollars flowing out of your accounts automatically like bleeding from wounds you didn't know you had.

Death by a thousand cuts. Each one too small to notice until you add them all up and realize you've been bled dry paying for conveniences that made your life worse while promising to make it better. Each subscription is a small surrender of independence, a tiny agreement to let someone else control some aspect of your life in exchange for not having to think about it or take responsibility for it.

But not thinking about things doesn't make them disappear. It just makes you vulnerable to people who are thinking about how to extract money from people who aren't paying attention. The subscription economy is designed to exploit human psychology, to make temporary access feel like ownership while ensuring that you never actually own anything.

Service Dependencies: Outsourcing Your Competence

Service dependencies were harder to cut because they exposed how little we actually knew. Convenience had crippled us. Learning to live without it felt like learning to walk again. One time, we hired a company to mow the lawn at our so-called dream home. They hacked it to pieces like they were paid to ruin it. After that, I did it myself. Because when you do things yourself, they get done right. You don't have to pay someone to pretend they care. They don't. Nobody gives a fuck about your life but you.

But now on our farm, I don't mow hardly anything aside from small paths so we can leave it wild for nature, wildlife, flowers, and whatever wants to grow without human interference. Most lawn services, cleaning services, and maintenance services exist because people don't have time to do things themselves. Time is what you get back when you stop working jobs that steal your time to pay for things you could do yourself if you had the time that the job stole.

If you control your own time, you can do most of these things better and cheaper yourself. You just have to be willing to learn and accept that it might not look perfect while you're learning. Competence is earned through practice rather than purchased through payments to experts who might not be expert at anything except extracting money from people who think they're incompetent.

The Financial Chains That Bind

One of the most difficult cords to cut were the financial ones, the chains made of money that felt like security but were really just sophisticated traps designed to keep us working until we died.

Mortgages and car payments had been such a constant part of our lives that we didn't realize how much mental energy they consumed until they were gone, like removing tumors we'd learned to think of as normal parts of our bodies.

Every month, thousands of dollars flowed out of our accounts to pay for things we'd already received. It was like being charged for the privilege of existing in a world that we'd helped build with our labor but didn't own.

We still have bank accounts, but we don't put all our eggs in one basket because baskets controlled by other people can be dropped or stolen or set on fire whenever it serves their interests. We spread money across multiple institutions and keep some physical cash on hand because digital money can disappear with the flip of a switch. Those switches are controlled by people who don't care if we live or die as long as we don't cause problems for their systems.

We don't trust any single system or individual to protect our access to our own money because systems are designed to serve the people who own them, not the people who depend on them. When those systems fail, they fail upward. The people who own them get richer while the people who depend on them get poorer.

The Liberation of Debt Elimination

The mortgage cord was the most liberating to cut, like removing a noose that had been tightening around our necks so slowly we'd learned to think of suffocation as normal breathing. Nobody can explain the feeling of never having a mortgage until you've paid rent or a mortgage to someone and then all of a sudden you don't have to anymore. You keep your money like keeping your own blood instead of donating it regularly to vampires who call themselves lenders.

Every month, instead of sending thousands to a bank that sees you as a risk to be managed and a resource to be exploited, you get to decide how that money is used. Psychological freedom is even more valuable than financial freedom, because it lets you think clearly about what actually matters instead of constantly calculating how to afford things you don't want for a life you don't even want to live.

The mortgage wasn't just a financial obligation. It was a psychological prison that kept us working jobs we hated to pay for houses we didn't want in neighborhoods full of people we couldn't stand, living lives that slowly killed us while we pretended to be successful. The mortgage was the chain that connected everything else. The job. The commute. The car payments. The insurance. The utilities. The maintenance. The taxes. The HOA fees. The endless expenses that come with owning property in a system built to extract maximum profit from basic human needs.

The Hardest Cut: Family Ties

But the hardest cord to cut was familial ties to those who didn't agree with our choices. People who saw our escape as a judgment of their captivity, and our freedom as an attack on their commitment to suffering. Some people will never understand why you'd walk away from financial security, career advancement, or social status, because understanding would require them to admit they've been wasting their lives pursuing things that don't matter while ignoring things that do.

My mother could not comprehend why we'd liquidate everything to live on a farm in West Virginia like we'd chosen to become homeless by choice. She cut us off like we'd died instead of choosing to live. The choice between family approval and personal freedom isn't really a choice when family approval requires you to keep dying slowly to make other people comfortable with their own slow death.

You can't live your life for other people's comfort, especially when their comfort depends on your suffering. When their sense of normalcy requires you to keep participating in systems that are actively destroying you. Some relationships can't survive honesty about what life is really like, or what choices are truly available to people who are willing to admit the system is designed to kill them.

Infrastructure Independence: Choosing Your Dependencies

Infrastructure independence is ongoing, like recovering from an addiction that the whole culture is designed to enable and encourage. We still depend on the electrical grid, though we're upgrading to solar so we can generate our own power instead of buying it from companies that can raise prices or cut service whenever they want more profit or more control. We still depend on municipal water systems in some ways, though we're working toward complete water independence so we can drink clean water without asking permission or paying fees to people who profit from basic human needs.

We still depend on internet service for our business, though we have backup options because depending on any single point of failure is like building your house on quicksand. We often get a chuckle and are a bit confused when we see other businesses post about their internet being down or the phones being down. Phones? Landlines?

Can't accept credit cards? How have businesses become so shitty at extracting profit from people that they can't even maintain their own systems anymore. The goal isn't to cut every cord immediately because cutting all cords at once is like jumping out of an airplane without a parachute. Dramatic but ultimately counterproductive.

The goal is to identify dependencies and reduce them systematically. Each cord you cut makes you more resilient and less vulnerable to other people's decisions and failures. More capable of surviving when systems break down, and less dependent on systems that are built to fail when failure benefits the people who own the replacements.

Starting the Cutting Process

Some cords are worth keeping if they truly serve your purposes and values, but most dependencies are just habits disguised as necessities. They become routines that feel normal only because everyone else is following the same patterns, slowly dying from the same causes. Most subscriptions pay for convenience you don't actually need. Most services exist to solve problems you could handle yourself with a little effort and a willingness to learn.

Start with the obvious ones like removing the most visible parasites first. Cancel subscriptions you don't use or use so rarely that you'd forgotten you were paying for them. Eliminate services you can replace with your own effort without significant hardship or risk. Reduce your dependence on systems that don't align with your values or serve your actual needs.

Then work on the harder ones, like performing surgery on yourself while other people tell you it's dangerous to remove tumors they've learned to think of as vital organs. Joking, obviously. Reduce debt that ties you to jobs you don't want but feel like you need, just to service obligations that benefit other people's interests. Develop skills that reduce your reliance on professional services that charge premium prices for basic competence. Build relationships that aren't based on financial transactions, but on mutual aid and shared values. Notice I said values, not religion. These are not interchangeable. Values are what you actually do when no one's watching. Religion is what you say you believe when everyone is.

The goal isn't isolation or paranoia. The goal is choice, the freedom to make decisions based on your values rather than your obligations to systems that see you as a resource to be consumed. When you depend on fewer external systems, you have more options. When fewer people control aspects of your life, you have more freedom to make decisions that serve your interests instead of serving the interests of people who profit from your dependence.

Every cord you cut is a vote for your independence, every dependency you eliminate is a step toward freedom from systems that are designed to keep you trapped and bleeding. Start small, but start. The system only has power over you as long as you need it more than it needs you, and it needs you a lot more than you've been told.

Part III: Living the Afterlife

Chapter 11: Our First Farm Animals

Our first farm animals were rabbits because we wanted a steady supply of clean meat that could be produced quickly and efficiently without depending on industrial agriculture systems that treat animals like manufacturing inputs. The same agricultural systems that treat consumers like revenue streams who aren't supposed to ask questions about production methods or animal welfare standards.

Want to know what food we feed our rabbits? I'll tell you the brand and why we chose it. I don't care what your opinion on the food is, because I don't make decisions in a vacuum. I know my rabbits are healthy. My bucks perform regularly and treat our does with care, not aggression. Our does are sweet and nurturing mothers that raise kits that easily grow to over three pounds dressed weight, and all of our does carry large litters. We treat every one of our animals with the utmost respect, and in turn they provide us with the best tasting food we've ever eaten. Better food than you'll ever eat unless you buy it from us. Even if you do buy food from us, you'll likely never have one of our rabbits that was alive 30 minutes before we cooked it. Truly the freshest meat you can get your hands on. Oh, you hunted rabbits? We raise and eat domesticated meat rabbits, and there is a big fucking difference. We'll never go hungry again, and the community is lucky we share any of it with them.

Initially we bought six New Zealand rabbits: two bucks and four does. All were thirteen weeks old. They were old enough to transport safely but young enough to adapt to new environments and handling routines without developing behavioral problems from disrupted territory or social hierarchies. We raised them from young rabbits to become meat producers, which sounds clinical but accurately describes the purpose they serve in our food production system.

One of the original bucks was culled later because it had temperament issues that made it unsuitable for breeding, which is a polite way of saying it was aggressive, difficult to handle, and showed early signs of poor behavior toward our does. It would have passed those traits to offspring, creating ongoing management problems for people like us who need to handle animals regularly for feeding, health checks, and processing. We never bred it, because breeding problem animals produces more problem animals. And we weren't running an animal rescue operation for rabbits with personality disorders.

We now have over forty rabbits at any given time at various life stages. That sounds like a lot until you consider that rabbits produce large litters frequently and grow to processing weight quickly. Maintaining breeding stock means keeping multiple generations at once to ensure steady production. Seasonal gaps would leave us without meat, requiring us to buy over-priced grocery store protein that tastes like complete dog shit, and is more expensive than what we raise ourselves.

Both of us had roots in country life long before we started farming. Alexys spent her childhood summers trapping gophers on her grandparents' land in North Dakota. I spent weekends and summers on my grandpa's sheep farm, helping out during the first eighteen years of my life until I left for college. We weren't complete novices who thought animals were pets that occasionally provided food. These were livestock. Their purpose was food. Understanding the difference is crucial for anyone who wants to raise meat without forming emotional attachments that get in the way of making decisions about which animals stay and which animals get processed. Those decisions need to be based on practical factors like health, behavior, and productivity, not on how much you like a particular rabbit.

The Learning Curve That Nearly Broke Us

We started researching rabbits online because the internet holds more information about rabbit husbandry than most people could learn in years of practical experience. But internet knowledge doesn't replace hands-on work with live animals, each with their own health needs, personalities, and behavioral quirks that can't be predicted or managed through theory alone.

The first week was overwhelming. We had never owned rabbits as pets, let alone raised them for meat production. That requires a different mindset and different management practices than keeping animals for companionship or entertainment. Meat production demands objective evaluation of performance, health, feed conversion, and processing schedules that have to be followed.

Our first major mistake was trying a colony setup, which is where multiple rabbits share a common living space rather than being separated into individual cages or hutches. At first, the colony model seemed appealing. It looked more natural, required less infrastructure, and promised a more hands-off approach to daily care. But what looked simple on the surface quickly became a management nightmare.

Without individual enclosures, it becomes extremely difficult to monitor feed intake, track health issues, manage breeding schedules, or identify problem behavior. Illness or aggression can go unnoticed until it spreads across the herd, and even routine checks become guesswork when you can't isolate which rabbit is struggling. There's no way to be sure who's eating, who's declining, or who needs to be pulled for treatment.

Colony setups also make it harder to provide targeted care. You can't easily separate animals that need individual attention, isolate those showing early signs of illness, or control breeding timing to align with production needs. Catching a specific rabbit for processing or health checks disrupts the entire herd and creates unnecessary stress, which directly affects feed conversion and weight gain in the remaining animals.

Individual housing requires more up-front investment in cages, feeders, and layout planning. But that investment pays off in control, predictability, and better outcomes for both the animals and the people caring for them. You can monitor each rabbit closely, make adjustments as needed, and maintain a healthier, more stable environment from top to bottom.

No animals died or were injured during our learning period. We researched extensively before purchasing living beings on impulse, which distinguishes responsible livestock owners from people who buy animals without understanding their needs. Those animals often get neglected due to ignorance or are abandoned when care becomes inconvenient, expensive, or time-consuming.

Animal deaths due to preventable causes represent failures of planning and preparation. They could have been avoided through better research and infrastructure development before bringing animals onto the property.

We wouldn't allow animals to die from preventable causes because animal welfare is both an ethical responsibility and an economic necessity for people who depend on animals for food production. Dead animals don't produce meat, and sick animals don't convert feed efficiently. Maintaining animal health serves both moral and practical purposes that align the interests of the animals with the interests of the people who depend on them for survival and income.

Building Everything Because Prefab is Garbage

We build all animal housing ourselves because commercial animal shelters are expensive garbage that breaks quickly and often causes harm or death to animals that depend on secure, weather-resistant housing. Housing that protects them from predators, extreme temperatures, and environmental hazards. Hazards that can injure or kill animals whose owners trusted manufacturers who prioritize profit margins over structural integrity and long-term durability.

We built hutches for rabbits and quail, chicken coops and runs, and various other animal housing structures, using materials and designs that prioritized functionality and durability over appearance and convenience. If you aren't building your own animal shelters, you're pissing money away on cheap shit that will break and likely cause harm or death to animals that trusted you to provide adequate protection from environmental threats and predator attacks.

We built everything except our first chicken coop. That's one reason why we never purchase prefab animal shelters anymore. The commercial coop was poorly constructed with materials that deteriorated quickly, fasteners that failed under normal use, and design flaws that created safety hazards and management difficulties. Those difficulties wouldn't have existed if we had built the structure ourselves using appropriate materials and construction methods.

Commercial animal housing is designed by people who don't raise animals, and built by people who don't use the products they manufacture. The designs prioritize manufacturing efficiency and shipping convenience over animal safety and user functionality. Marketing photos show attractive structures that look adequate for people who don't understand animal behavior and housing requirements. But actual use reveals design flaws and construction shortcuts that become apparent only after animals have been injured or killed by preventable structural failures.

Building your own animal housing costs more initially in time and materials. But it produces structures that last longer, work better, and can be modified or repaired by people who understand the construction methods and materials. The reason they now understand them is because they built the structures themselves using tools and techniques they can replicate when maintenance or modifications become necessary.

The Scaling Decision That Wasn't Really a Decision

We scaled up immediately after our first rabbits. Scaling up is necessary rather than optional when you're producing animals that reach sexual maturity quickly and produce large litters that must be managed appropriately. It is a priority that we manage our animals appropriately to prevent overpopulation problems that can overwhelm housing capacity and feed budgets if breeding isn't planned and controlled according to processing schedules and market demand.

After our first rabbits, we bought thirty quail. Now, less than a year later, we have over one hundred fifty quail at various stages of development, from day-old chicks to mature breeding birds. They produce eggs for eating and hatching, depending on seasonal demand and production goals that change throughout the year based on customer requests and our own consumption needs.

Most of our quail are laying hens that produce eggs for eating, but we also keep roosters in some hutches for fertilized eggs that we incubate and hatch for the next generation of meat birds or laying hens that replace older birds that stop laying efficiently. Some may develop health problems that make them unsuitable for continued production. Always cull your bad breeding stock. Got an asshole rooster quail that is overbreeding or scalping your hens? Pull him out and have him for dinner that night. Don't wait for that rooster to start acting right because you'll come out one morning to feed your covey and find a dead hen or a couple. I haven't, but that's because these are livestock, and we don't wait for them to correct themselves. You must take action. You must cull that rooster. You must be able to make decisions, because if you can't, you can never leave the systems that bind you.

Maintaining breeding flocks and coveys requires keeping multiple generations simultaneously to ensure genetic diversity and continuous production capacity that isn't disrupted by disease outbreaks or other problems that can affect entire age groups simultaneously.

We didn't feel confident about breeding rabbits when we started, but we didn't have a choice because sexually mature rabbits will breed whether you feel ready or not. Uncontrolled breeding produces more animals than you can house or feed or process appropriately if you don't have systems in place to handle increased production that results from successful breeding programs.

This is why our rabbits have individual housing. Even if you don't breed rabbits who are kept individually, not breeding produces less than desirable results as well. Does fake pregnancies and start building nests for rabbit kits that will never be born. Bucks spray piss ALL OVER THE PLACE.

All animals start out in our house so they can get used to our pets and human interaction that makes them easier to handle throughout their lives. Our chickens are brooded indoors. Our quail are raised indoors until they're large enough to survive outdoors. Our main breeding rabbits lived indoors with us initially to establish trust and handling tolerance that makes management easier when animals need individual attention for health monitoring or breeding management.

Our pets don't bother the livestock because they've been exposed to farm animals regularly since we started raising them, and they understand that farm animals are different from wild animals that might be appropriate targets for chasing or hunting behavior. Dogs that grow up around livestock learn to distinguish between animals that belong on the property. Animals that represent threats or opportunities for predatory behavior must be controlled to protect valuable livestock from domestic animals that might injure or kill them if not properly trained and supervised.

The Psychology of Processing Your Own Food

The transition from buying meat to producing meat requires a psychological shift. Most people aren't prepared for it because they've been conditioned to think of meat as something that simply appears in grocery stores, not as part of an animal that lived and died to feed them.

Our animals are nutrition for us, and we are willing to take responsibility for the entire food production process, from birth to processing to consumption. Not ignore the worst parts. People who ignore the worst parts shift that burden onto people like me, who will kill meat for others while they feel good about supporting small-scale agriculture but not assuming any of the responsibilities that go along with it.

Most people prefer to remain ignorant about how their food is produced because knowledge creates responsibility, and responsibility requires making decisions based on values rather than convenience. It's easier to buy meat from sources you don't investigate than to raise animals yourself and take responsibility for their welfare throughout their lives, and for ending their lives humanely when they reach processing weight or age.

Processing your own animals requires accepting that death is part of life, and that eating meat requires someone to kill animals regardless of whether you do it yourself or pay other people to do it for you in industrial facilities. Those facilities prioritize efficiency over animal welfare and prioritize profit over humane treatment of animals. They and we provide food for people who don't want to think about the violence that makes their meals possible.

Raising your own meat animals allows you to control every aspect of their lives and deaths. You ensure that the animal receives appropriate care and nutrition, and that their deaths are as quick and stress-free as possible given the inherent realities of converting living animals into food for human consumption. Industrial meat production prioritizes cost reduction over animal welfare, so animals raised in factory farms often live in conditions that would be considered torture if applied to animals that people consider pets rather than livestock.

The first time you process an animal you raised yourself, you understand the real cost of meat in ways that grocery store purchases never teach. You understand the responsibility of taking life to sustain life, the work required to convert living animals into food, and the satisfaction of knowing exactly how your food was produced and what the animal experienced throughout its life under your care and management.

From that point forward, store-bought meat tastes different because you know it represents animals that lived and died in conditions you probably wouldn't accept for animals under your direct care. Once you've produced your own meat according to your own standards, commercial meat seems like a compromise that serves convenience more than conscience, profit more than principle, and efficiency more than ethics.

The Numbers Game That Actually Matters

Maintaining forty rabbits at various life stages sounds like a lot of work, until you consider that forty rabbits can produce more meat in a year than most families consume. Rabbits accomplish this using feed that costs less than grocery store meat, while providing protein that's fresher, cleaner, and more nutritious than anything available through commercial sources. Those sources prioritize shelf life and shipping convenience over nutritional value and flavor.

Rabbits convert feed to meat more efficiently than most other livestock. They reproduce quickly, reach processing weight in a few months, and don't require extensive infrastructure or specialized equipment. All of that makes other livestock impractical for small-scale producers who want to produce their own meat without investing in facilities and equipment that cost more than the meat is worth over reasonable payback periods.

Quail produce eggs and meat in even smaller spaces with even less infrastructure. They reach maturity faster than chickens while requiring less feed per pound of production. A hundred quail can produce eggs for a family while taking up less space than a few chickens. Quail eggs are considered delicacies that sell for premium prices to customers who appreciate their nutritional value and unique flavor characteristics.

The numbers work when you calculate the cost of feed versus the cost of store-bought meat and eggs. But the real value is in the quality difference, and in the knowledge that your food was produced according to your standards rather than commercial standards that prioritize profit over everything else that matters for nutrition, animal welfare, and environmental impact.

Self-sufficiency in meat production eliminates dependence on supply chains that can be disrupted by weather, transportation problems, economic instability, or political conflicts that affect food distribution systems. When you produce your own meat, you're not vulnerable to shortages, price increases, or quality problems that affect commercial meat supplies that most people depend on for protein needs. Needs that can't be met through other sources when industrial systems fail.

The learning curve for animal husbandry is steep initially, but levels off quickly once you understand basic principles of animal behavior, nutrition, housing, and health management that apply across different species. These differences contain minor variations that can be learned through experience and observation, rather than formal education or professional training that costs money and takes time that could be spent gaining practical experience with actual animals.

Starting with animals was the beginning of real food security, rather than the illusion of food security that comes from having money to buy food that other people produce according to standards you don't control. Using methods you don't understand. Produced in disgusting facilities you've never seen, with ingredients you can't pronounce, for purposes that serve profit more than nutrition.

Chapter 12: Our New Economy

The transition from collecting a paycheck to creating goods fundamentally changed how we think about income and security. Instead of depending on someone else's decision about what our time is worth, we now create products that people actually want and need. Just because we're a farm doesn't mean we're seasonal either. While most people think agriculture shuts down in winter, our operation runs year-round because we've built it to serve customers regardless of the season.

People from all around the country buy rabbit manure year-round. Lucky for us, our rabbits shit year-round as well. It's not always winter in Alaska. It's not always winter in Minnesota. There's always someone, somewhere, starting a garden who wants to buy our triple sifted premium rabbit manure.

The Heritage Revival

Our customers vary, but they're usually older people drawn by nostalgia. Rabbit and quail are heritage meats that are fading away in this part of the country, and we're helping to revive them. These are foods that their Appalachian grandparents raised and ate, proteins that were common on American tables before industrial agriculture made them seem exotic. We also get foodies who appreciate the quality difference, and bakers who specifically value our quail eggs for their rich yolks and superior binding properties. These customers understand that what we're producing isn't just food. It's food the way it used to be made, before factory farming and mass production stripped away flavor and nutrition in favor of convenience and profit. Customers who don't get it only buy our chicken eggs and nothing else. They're looking for a cheaper alternative to store-bought eggs, especially during an egg shortage, or they want to post on social media about the variously colored and sized chicken eggs they purchased from the small farm stand outside of town. It's always interesting seeing people claim ownership of us like that. They're transactional customers who don't understand or care about the larger picture of self-sufficiency and food security. Customers who do get it buy everything we produce and ask questions about how we raise the animals, what we feed them, how we process them. They want to learn. They're interested in the philosophy behind what we're doing, not just the products. These are the customers who become repeat buyers and refer their friends and family.

A Day in the Real Economy

I wake up at 4am and take Moira, our Bernese mountain dog, out to the bathroom, then move directly into feeding the rabbits. Each rabbit gets checked individually for health, water, food levels. Then it's on to the quail, making sure they have everything they need. Finally, Alexys, Clemmy our lab, Moira, and myself let the chickens out of their coops and give them their food and scratch grains. Then it's on to projects. Watering crops, building a new hutch, maintaining or fixing equipment, harvesting vegetables. All of this with our dogs by my side. They're my constant companions and supervisors, following me from task to task throughout the day. We break for lunch and chill with our dogs under our big black walnut tree. No rushing through a sandwich at a desk or eating while checking emails. Actual rest, actual conversation, actual peace in the middle of the workday. The afternoon repeats the morning routine. Animals are rewatered, refed, checked on. Then back to whatever project I'm working on. Around 5pm, the dogs get fed and work starts winding down. We have dinner together at a table with each other every night. No TV or distractions. Just us, talking about the day, planning for tomorrow, enjoying food we raised ourselves. During the evening, we do final rounds, making sure everything is locked up and the animals have the resources they need to make it through the night. Then we're done. Work ends when work ends, not when a clock says it should or when someone else decides we've done enough.

The Corporate Comparison

Corporate life was to wake up, chug coffee, drive to work or, if working from home, start working early to get tasks done before calls started. Then the day was a blur of meetings, interruptions, and usually little food or downtime. Head home, bitch about work, wind down with copious amounts of wine or whiskey because that's classy, right? We don't drink anymore by the way. Go to bed and wake up and do it again. The contrast is stark. My corporate days were filled with time but empty of purpose. My farm days are filled with purpose and the time takes care of itself. Corporate work was abstract: meetings about meetings, reports that nobody read, projects that got canceled or changed constantly. Farm work is concrete. Animals are fed or they're not, plants are watered or they die, repairs are made or things break. In corporate life, success was measured by other people's metrics: performance reviews, quarterly goals, revenue targets that had nothing to do with the actual work being done. In farm life, success is measured by results you can see and touch and eat. The animals are healthy. The plants are growing. The freezer is full of meat we raised ourselves.

The Balancing Act

How I decide what to work on each day depends on all of the above: animal needs, weather, seasons, and our own priorities. It's a balancing act, but not in the traditional sense where you're constantly stressed about conflicting demands. It's easy to swap a bad weather day for doing dishes or building something inside a shed. The animals come first, always. They depend on us completely, and their needs are non-negotiable. But beyond that, the work flows naturally based on what's most urgent, what the weather allows, and what fits with our energy levels and interests. If it's raining, we work on indoor projects. If it's hot, we work in the shade or do tasks that require less physical exertion. If we're harvesting, everything else waits. If it's breeding season, we focus on the animals. The rhythm makes sense in ways that corporate schedules never did. But sometimes I'm forced to work in really shitty weather. Unfreezing animal waterers in bitter cold. Processing animals in extreme heat. It's got to be done, but I love every minute of it. And I wouldn't trade it for a day filled with shitty zoom calls and even shittier people.

Real Recession-Proofing

We couldn't sustain ourselves forever on what we produce now, but that's the goal we're working toward. What I do know is we can sustain ourselves a lot longer than the vast majority of people could, which feels good. We'll never be the ones at the grocery store looking like assholes because eggs are sold out and cost $10 per dozen. If everything collapsed tomorrow (supply chains, financial systems, electrical grids), we'd have meat in the freezer, eggs from our chickens and quail, vegetables from our garden, and beautiful flowers to add color and texture to our day. We'd have the skills to process our own animals, preserve our own food, and maintain our own infrastructure. Most people have three days of food in their houses and no idea how to produce more. Most people depend completely on systems they don't control and can't fix when they break. When those systems fail (not if, when), they'll be helpless and desperate. The new economy is highly recession-proof because we make our own food for just the cost of feed. People who make $8 per hour might not be able to purchase a dozen eggs with that $8, but I can feed a lot of rabbits for a month with that same eight dollars. Our food security doesn't depend on our income or anyone else's economic decisions. Feed costs don't fluctuate too terribly. They especially don't if you grow your own animal feed.

The Feeling of Real Income

You'll never know the feeling of making money from something you created until you actually do it. It's a level of accomplishment that is unmatched. Every dollar we earn comes directly from value we created with our own hands and knowledge and effort. When someone buys our eggs, they're buying the result of us caring for chickens and quail every day, providing them with good food and clean water and safe housing. When someone buys our meat, they're buying protein we raised from birth, processed ourselves, and packaged with our own standards of quality and cleanliness. There's no middleman taking a cut. No corporate overhead reducing our compensation. No performance reviews or office politics determining our income. No layoffs or reorganizations threatening our livelihood. The money we make is directly connected to the value we provide. In corporate life, your paycheck is disconnected from your actual contribution. You might work incredibly hard on projects that get canceled, or produce excellent results that nobody notices, or coast through periods where you accomplish nothing meaningful, and the paycheck stays the same. The compensation has no relationship to the value created. In the real economy, everything is connected. Better animals produce better products. Better products attract better customers. Better customers pay better prices and refer more business. The feedback loop is immediate and honest.

Direct Relationships, Direct Results

Local people mainly buy our food and flowers. We don't sell to restaurants or participate in farmers markets. We attempted to work with restaurants, but they just upsell our products and make them unaffordable for regular people. Farmers markets don't work for us. They're like neighborhood garage sales where everyone is competing for an individual dollar. The odds aren't good. Instead, we make good products that people want to come to us for. People drive to our farm to buy eggs, meat, and flowers. They place orders through our website. They refer their friends and family. We build relationships with customers who value what we produce and how we produce it. This direct relationship eliminates the markup and manipulation that happens in traditional retail. Our customers know exactly where their food comes from, how it was raised, and who raised it. We know exactly who's buying our products and what they value about them.

Sustainable, Not Scalable

We invest everything we make in ourselves. Whether that's farm infrastructure, equipment that makes us more productive, or knowledge that makes us more capable. It's all for us. We don't invest in other people's companies or trust our future to markets we can't control. Our time is purposeful and for us now, not a corporation. We work the same hours we worked in corporate jobs (4am to 10pm or midnight), but we're always home. We always eat dinner together. We always see our pets. We eat good, clean food instead of grabbing shitty coffee for a long shittier commute by car or bus on the way to work. A good day now is every day. We took care of everything that needed taking care of. Challenges come and go, but those don't create bad days anymore. Problems are puzzles to solve, not crises that threaten our identity or security. We price everything fairly and below what you might get elsewhere. Our eggs are cheaper, our meat is cheaper, our flowers are cheaper. Because we do it all ourselves. Alexys picks and hand-ties the flowers herself. We raise and process the meat. We don't pay anyone for labor we can provide ourselves. This isn't a scalable business model in the traditional sense. We can't hire employees and expand locations and franchise the concept. But it doesn't need to scale. It needs to sustain us, and it does that effectively while giving us the life we actually wanted to live.

Chapter 13: Slowness as an Exit

We aren't protesting anything because protesting implies that we still believe the system can be reformed. That the people who profit from our suffering can't be convinced to stop profiting from our suffering if we just ask nicely enough or make enough noise or vote for the right candidates who promise to fix problems they helped create. We're done getting scraps in return, like beggars asking for better treatment from masters who see us as livestock to be managed rather than human beings to be served.

But living slowly in a speed-obsessed culture is inherently political, whether we intend it that way or not. It's like breathing clean air in a polluted city or eating real food in a world full of synthetic substitutes. It's a rejection of the assumptions that faster is better, that busy means productive, that optimization and efficiency and maximum output are more important than peace and health and relationships and meaning.

The pace shift isn't about working fewer hours because working fewer hours while trapped in systems designed to consume you just means being consumed more slowly, like dying of cancer over years instead of months. I still work long days, but the quality of those hours is completely different, like the difference between being tortured and being alive.

Natural Rhythms vs. Corporate Schedules

The land and animals determine our schedule, plus our own needs and desires, instead of schedules imposed by people who profit from our compliance with rhythms that serve their interests rather than ours. We have responsibilities to the livestock because they depend on us completely. Beyond that, our time belongs to us like our bodies belong to us instead of being rented to corporations that use them until they break and then replace them with younger models.

We think in seasons rather than quarters or fiscal years like living in the real world instead of the artificial world. A world created by people who measure success in numbers that have no relationship to human wellbeing or environmental health, or really anything that actually matters for survival and happiness. Spring is for planting and breeding. Summer is for growing and maintenance. Fall is for harvesting. Winter is for planning and recovery. These rhythms make sense in ways that quarterly earnings reports never did.

We don't really care if it's Tuesday or a holiday because it doesn't matter anymore when your work has meaning and your time has value. It matters when your life has purpose beyond generating profit for people who see you as a cost center to be minimized rather than a human being to be valued. The animals need care every day regardless of what the calendar says. The garden grows on its own schedule regardless of what meetings are scheduled or what deadlines are approaching.

Moving at the day's pace instead of artificial urgency allows us to accomplish much more physically while expending less mental energy. Before, we kept up with manufactured deadlines and artificial crises designed to keep us busy and anxious and compliant. Now we respond to real needs and genuine priorities that actually matter for our survival and wellbeing.

Urgency still happens when animals get injured or heat waves hit or equipment breaks down, but we're well-equipped to handle those stressful moments. We are equipped to handle them because they're real problems with real solutions that can be implemented by real actions, not manufactured crises designed to keep us distracted. Distracted from the fact that most of our work serves no useful purpose except generating profit for people who contribute nothing of value to the world.

Information Diet as Survival Strategy

We don't watch or follow the news because news is designed to keep people anxious, reactive, and consuming information. It makes people feel informed while actually making them less capable of understanding what's really happening, and why it matters and what they can do about it. Daily stock markets, what someone tweeted, or whichever political theater stunt is going on: we don't partake in it because it controlled our lives before and now we don't allow it in like refusing to let poison into our bodies.

The constant stream of information keeps us emotionally invested in problems we can't solve. It pulls us away from problems we can solve, in places where we have actual influence, with people who actually matter for our wellbeing and survival.

News urgency differs from real urgency. One creates fake emergencies designed to generate clicks and advertising revenue. The other presents actual emergencies that threaten things we care about, things we can actually impact through our actions.

When animals get injured, we act. When heat waves threaten our crops, we respond. When equipment breaks down, we fix it. We handle these situations immediately and effectively because they're real problems affecting our real lives. They can be solved through real actions we can take ourselves.

We don't create artificial stress by consuming information about problems that exist primarily to generate engagement and advertising revenue. Media companies profit from keeping people anxious, angry, and helpless. They provide no useful information about how to solve problems, improve situations, or build alternatives to the systems that create the problems in the first place.

Slowness as Resistance to Acceleration Culture

Slowness as an exit means refusing to participate in the acceleration that keeps people too busy to think, too distracted to question. That same acceleration keeps people too exhausted to change anything meaningful about their lives or the systems that control those lives. It means choosing deliberation over reaction, contemplation over consumption, presence over productivity optimization.

We wash dishes by hand because it uses less water and gives us control over the process, but also because it forces us to slow down and pay attention to what we're doing. It's easy to find yourself rushing through tasks while thinking about the next task and the task after that in an endless chain of busyness that prevents us from ever being fully present at any moment.

We buy CDs because we want to own music instead of renting access to it, but also because choosing physical media requires intention and deliberation. Impulse streaming keeps us consuming content without ever fully experiencing or appreciating any of it.

We line-dry laundry because it's gentler on clothes and doesn't require electricity, but also because it connects us to natural rhythms and forces us to work with weather instead of against it.

These aren't revolutionary acts in themselves. They're choices that prioritize quality over convenience, intention over impulse, presence over productivity. They're small acts of resistance against a culture that treats every moment as an opportunity to optimize efficiency rather than experience meaning.

Redefining Productivity as Human Flourishing

The protest is in the refusal to optimize every moment for productivity or efficiency or maximum output as defined by people who profit from our labor rather than our wellbeing. Sometimes we sit and watch the animals because watching animals is pleasant and peaceful and connects us to the living world instead of the artificial world. Where screens are king and the world is filled to the brim with notifications and other demands for our attention from people who want to sell us things we don't need.

Sometimes we take long walks without destinations because walking is good for our bodies and our minds. It gives us time to think without being interrupted by phones or emails or other people's agendas for our time and attention. Sometimes we eat meals without doing anything else simultaneously because food tastes better when you pay attention to it and meals are more satisfying when they're not just fuel consumed between tasks.

These moments of presence and attention are radical in a culture that treats them as waste, that measures the value of time by how much output it generates rather than how much peace or joy or connection or understanding it provides to the people experiencing it.

Productivity gets redefined when you control your own time and work toward your own goals instead of goals imposed by people who profit from your labor. I've already established that they contribute nothing of value to your life or the world. A good day isn't measured by tasks completed or emails answered or meetings attended or revenue generated for people who see you as a cost center to be minimized.

A good day is when the animals are healthy and the plants are growing, and we spend time together doing work that matters for our survival and wellbeing and future security. A good day is when we move closer to our goals without sacrificing our values or our health or our relationships for artificial deadlines imposed by corporations that don't care about our wellbeing.

The Revolutionary Act of Contentment

The slowness isn't laziness or lack of ambition or failure to understand the importance of hustle culture and maximum productivity optimization. It's intentionality born from the recognition that busy isn't the same as productive, and productive isn't the same as meaningful, and meaningful isn't the same as valuable, according to people who profit from our busyness.

Slowness is choosing what deserves our time and attention instead of letting other people's priorities colonize our schedules and our thoughts and our energy. It's recognizing that the urgent rarely overlaps with the important. Most of what feels urgent is designed to feel urgent to generate compliance with systems that serve other people's interests rather than our own.

The revolution is in refusing to accept that our worth as human beings is measured by our output as workers, that our value is determined by our productivity rather than our humanity, and that our time belongs to anyone except ourselves.

Chapter 14: What You Keep, What You Burn

We didn't keep everything. That's the point. We didn't want to. We shed what could be shed, cut what could be cut, and burned what needed to be burned. But not everything can be discarded. Not everything should be. Some systems still serve a purpose, even if they're flawed. Some dependencies are necessary, not because we're weak or undisciplined, but because we're human. Because real life isn't a purity contest. It's a triage of tradeoffs.

And the first non-negotiable in our world is survival. Which means some things we kept, not out of ideology, but out of reality.

Healthcare we kept, obviously, because reality doesn't care about your philosophy when your body starts breaking down and you need medical intervention to survive. Alexys has MS, and she's my number one priority. She is more important than our independence, or our principles, or our desire to escape from systems that see us as profit centers rather than human beings.

We're middle-aged at this point. I'm 41, she's 36. Yes, middle-aged. We have a lot of life left, and we don't claim to have written the whole playbook yet. We haven't figured out everything for people trying to navigate serious medical conditions while maintaining independence from employment-based insurance systems. Systems that trap people in jobs they hate to access healthcare they need.

We currently have healthcare because we're not idiots who confuse ideology with survival strategy, and Alexys's care comes before everything else. The reason being is that everything else is meaningless if she's suffering or dying from lack of medical care.

We're not naive about medical realities or seduced by alternative medicine promises that sound appealing but don't actually work when you're dealing with autoimmune diseases. Diseases that can disable or kill you if you don't have access to treatments that require pharmaceutical companies and medical infrastructure that we can't replace with good intentions and herbal supplements.

Diet doesn't cure MS, but it can assist with some of the symptoms, and our diet is as clean as it comes. Real food from animals and plants we raise ourselves, without chemicals or additives. Without processing designed to maximize shelf life and profit margins while minimizing nutritional value. Alexys also stays moving on the farm, and the physical activity helps her manage the condition better than just sitting at a desk for eight hours a day while slowly dying of stress and inactivity.

But we're not delusional about what clean food and exercise can and can't do for serious medical conditions that require serious medical intervention. We aren't naive to the fact that diseases can't be cured by positive thinking or lifestyle changes alone.

Financial Infrastructure Without Financial Dependence

We kept bank accounts, but as I already mentioned, we don't put all our eggs in one basket, because baskets controlled by other people can be dropped or stolen or destroyed whenever it serves their interests or when systems fail. And they do fail. Fail due to incompetence or malice or simple entropy that affects all complex systems eventually.

Recent banking instability has proven that caution about financial institutions is warranted rather than paranoid. Spreading risk across multiple systems is smart rather than excessive. It's not a bad idea to keep some wealth in forms that can't be deleted by computer glitches or policy changes or executive decisions. That is prudent rather than primitive.

Critical thinking skills from our corporate jobs have carried over as those skills transfer directly to farm life, business operations, and problem-solving in situations where failure has real consequences. Not just outcomes that affect quarterly profit margins for people who don't do any of the actual work. We just apply those skills to problems we actually care about solving, instead of problems that exist primarily to justify the salaries of people who create more problems than they solve.

The ability to analyze situations, evaluate options, and implement solutions is valuable regardless of context. But it's more satisfying when you're solving problems that matter for your own survival and wellbeing. It is never very fruitful solving problems that exist to generate billable hours or justify organizational structures that serve no useful purpose except extracting profit from human labor.

What We Burned to Save Our Lives

What we burned was the debt that was slowly strangling us like a python that tightens its grip every time you exhale, making it harder to breathe until you suffocate while people around you applaud your "financial responsibility" and "investment in your future." Mortgages, car payments, credit dependencies, all of it went into the fire like burning the contracts that made us slaves to people who profit from our suffering.

The psychological freedom of not owing anyone anything is more valuable than any convenience those debts might have provided. It is more liberating than any lifestyle those payments were supposed to enable. When you don't owe money to banks or credit card companies or car dealers, you can't be threatened with repossession or foreclosure or credit damage for making choices that serve your interests rather than the lender's interests.

We burned corporate urgency and artificial deadlines that existed to keep us busy and anxious and compliant with systems that serve other people's priorities rather than our own. The manufactured crises were designed to make us feel important while keeping us distracted from the fact that most corporate work serves no useful purpose. It exists mainly to generate profit for people who contribute nothing of value to the world.

The pressure to respond immediately to emails and requests that weren't actually urgent, just labeled urgent to make people feel like their time and attention don't belong to them. The guilt about taking time for ourselves or prioritizing our own needs over the demands of people who see us as resources to be consumed rather than human beings to be respected.

We burned the news cycle addiction and political theater consumption that was designed to keep us emotionally invested in problems we can't solve involving people we don't know in places we can't affect. The daily outrage was designed to keep us feeling informed while actually making us less capable of understanding what's really happening and what we can actually do about problems that affect our real lives in our real communities.

The constant stream of information about things happening to people we don't know made us feel connected to important events while actually disconnecting us from the people and places and problems. People, places, problems that we can actually influence through our actions and choices and relationships.

Service Dependencies: Reclaiming Competence

We burned service dependencies wherever we could replace them with our own capabilities. We reclaimed skills that had been surgically removed from our brains to make room for dependencies that kept us helpless and profitable for people who charge premium prices for basic competence.

Lawn services, cleaning services, maintenance services. Services that were needed mainly because we didn't have time to do things ourselves when our time was owned by jobs that stole our days and energy. Jobs that left us too exhausted to maintain our own lives. Now we have time, because we control our time, and we've learned that doing things ourselves is often faster and better than coordinating with service providers who don't care about the results as much as we do.

We burned convenience subscriptions that were quietly draining money every month like financial vampires that feed on small amounts of blood so regularly that you don't notice you're being exsanguinated until you add up all the small cuts and realize you're bleeding to death from a thousand tiny wounds.

The Gray Areas: Strategic Dependencies

What we're still figuring out is healthcare long-term. Navigating serious medical conditions while maintaining independence from systems designed to trap you in employment-based insurance is risky. We're prepared for various scenarios, but we don't know what the medical system will look like in five or ten years. Costs keep rising, quality keeps declining, and access becomes more restricted to people who can afford premium prices for basic care.

Nobody can even claim to guess what healthcare might look like when the current system collapses under its own weight of bureaucracy.

We currently have grid electricity, but we're upgrading to solar for energy independence, so fuel shortages likely won't matter in the future when supply chains become more unreliable and energy costs become more volatile. Grid infrastructure is becoming more fragile due to deferred maintenance and increasing demand from systems designed to consume maximum resources while providing minimum service.

Food shortages aren't a concern since we produce our own meat, fruit, and vegetables without depending on industrial agriculture or global supply chains. Those can be more easily disrupted by weather, politics, economic instability, or simple incompetence from people who manage systems they don't understand for profit rather than resilience.

But medicine greatly impacts Alexys, so having access to it is crucial for her survival and quality of life, regardless of what it costs or what systems we have to interact with to access it. We can grow and raise most of what we need to live, but we can't manufacture prescription medications or perform medical procedures. Those require specialized knowledge and equipment that we can't replicate or replace with good intentions and internet research.

The Decision Framework: Values vs. Pragmatism

The gray areas are technologies and systems that provide genuine value, but come with risks or compromises that require careful evaluation rather than ideological rejection or uncritical acceptance. Internet access is essential for our business, but it connects us to systems we don't fully control. It can be monitored or restricted by people who don't share our values or priorities.

Email and websites provide platform independence compared to social media, but they still depend on infrastructure we don't own. They can be affected by policy changes. These platforms are vulnerable to technical failures or economic disruptions that we can't predict or prevent through our own actions.

We approach these dependencies strategically rather than ideologically, evaluating them based on whether they serve our actual needs rather than whether they conform to abstract principles. Principles that might not apply to our specific situation, goals, and constraints.

We use what serves our purposes while maintaining alternatives. We maintain exit strategies for when those systems fail or become hostile to our interests. We don't reject useful tools because they're imperfect, but we don't become completely dependent on tools we can't control or replace when they stop working or start working against us.

The goal isn't purity or total independence. Purity is a luxury that people struggling to survive can't afford. Total independence is impossible for people who want to maintain connections to other human beings, communities, and knowledge that exists outside their immediate experience.

The goal is reducing unnecessary dependencies while maintaining access to things that genuinely improve our lives or protect our health. Or they enable us to build relationships and share knowledge, and contribute to communities that share our values even if they don't share our lifestyle choices.

Some things are worth keeping even if they come with compromises, if the benefits outweigh the risks and the alternatives are worse than the problems we're trying to solve. Some things are worth burning even if they once provided real benefits, if the costs outweigh the benefits and the dependencies they create threaten more important values and goals.

The art is in distinguishing between the two and making conscious choices rather than just accepting whatever everyone else accepts. Don't just accept what everyone else accepts, just because it seems normal, especially when normal is slowly killing you and everyone around you mistakes familiar suffering for inevitable suffering.

Chapter 15: What Happens If You Win

Long-term success means maintaining what we've built until we don't have to anymore. When we can close our doors for good and just hide in these West Virginia hills. We'll play our version of the game on our terms, but nothing more.

We don't think about growth. We think about each other and our animals. Everything else is just noise. Scaling and expanding and building empires? That's the same trap we escaped from, with different branding.

Legacy doesn't concern us. We'll be dead, and we don't care what happens then. We built something for us on this planet. We were left with garbage to clean up from previous generations, and that's what we did. It's not our obligation to create a legacy when we don't care about fame or recognition or having our names remembered.

We control every aspect of our level of involvement with the outside world. Not the public, not our family, not anyone else. When something happens we don't like, we pull back and recede from society further, a reminder why everything outside these fences is problematic.

What's the point of living then, you might be asking? Alexys and I love each other more than anything in the world. Had we not found each other, I can't guarantee that I'd be around today. If something happens to her, I can't guarantee that I'll be around long after that.

That doesn't mean that if you haven't found someONE that keeps you on this fucked up earth, that someTHING won't. You may only need peace in the woods, a dog by your side, a cat in your lap. Find that one thing, and live for it. Protect it at all costs. Those are the only things worth living for.

If you haven't found something, look for it. Think you'll hate cats? I guarantee you won't. Allergic to them? Stop making excuses and get outside of your comfort zone and try something real. This is not real medical advice, by the way. Use common sense before anything. Don't go get a cat because some asshole in a book said you might like them just because he adores his cat.

The influence trap doesn't worry us. We're not trying to become experts or gurus or thought leaders. We're not building a personal brand or trying to monetize our lifestyle. We share what we've learned, but we don't pretend to have all the answers or to be qualified to solve other people's problems.

We're pretty proud of what we built. Smug about it even. There was a lot of dissatisfaction with myself in my previous life. Now, I built a life I can be smug and satisfied with.

We spent a lifetime chasing carrots at the end of sticks that were never there. Now we grow our own carrots.

Content and complacency might be the same thing, and that's fine. We've earned the right to be satisfied with what we have, instead of constantly striving for more. The endless pursuit of improvement and optimization and growth was killing us slowly. Contentment is the antidote.

We have a very close circle of people we interact with, but we rely on nobody but ourselves. You can't depend on other people. People die. People flake. People lose interest. People don't care. The only reliable support system is the one you build for yourself.

Healthcare is the one dependency we can't fully exit. We've built workarounds for almost everything else, but this is the system we still have to bow to, because not doing so could cost Alexys her life. Planning for that isn't theoretical. It's the math of risk, cost, and survival.

We don't pretend to have answers. We just prepare for the day the system collapses further or locks us out entirely. Until then, we keep one foot in their world and both eyes on the exits.

If lots of people followed our path, it would probably improve things for everyone. More local food production. Less dependence on fragile supply chains. Fewer people trapped in jobs they hate. More resilience in communities.

But people will create whatever version of reality they want, or what's convenient for them. It's like how people adapt diets for cheat days or create justifications for why something might be paleo or keto or vegan when it clearly isn't. People create bastardizations of reality all the time.

This path is ours, and it's not for anyone else to modify or improve or scale.

For us, winning looks like waking up every day and choosing how to spend our time based on our values and priorities, instead of other people's demands and expectations. It looks like eating food we raised and trusting where it came from. It looks like working hard on things that matter to us, instead of working hard to make other people wealthy.

Success is sustainability, not growth. It's maintaining what we've built without burning out or compromising the values that guided us here. It's being able to say no to opportunities that don't align with our goals, even when those opportunities look good to other people.

If we win, we get to keep living the life we designed, instead of the life we were sold. We get to stay home, stay together, and stay focused on what actually matters to us. Everything else is other people's problems to solve or ignore as they choose.

Chapter 16: Time Belongs to You Again

My sleep schedule is roughly the same as it was in corporate life. I start early and end late. But the fundamental difference is that I love what I do, and I don't have to deal with assholes all day who exist solely to justify their salaries by making other people's work more difficult and less productive. The hours are similar, but the quality of those hours is completely different. Like the difference between existing and living. Between surviving and thriving.

I stopped checking the time once I got busy enough with real work that time stopped mattering. Now it's sunrise and sunset, not corporate schedules built to control other people's hours.

Sitting in a cube or office staring at screens while pretending to be busy makes you do nothing but watch the clock like a prisoner counting down time until release. Fake work creates artificial awareness of time's passage that makes every minute feel like an hour, every hour feels like a day, and every day feel like a slow death. A slow death punctuated by brief moments of hope that tomorrow might be different, even though you know it won't be.

When you have real work that produces visible results and meaningful progress toward goals that matter for your survival and wellbeing, you don't care to glance at clocks or phones. You just work until tasks are completed, problems are solved, animals are fed, and projects are finished. No follow-up meetings required to discuss what was accomplished or plan what should be accomplished next, because the work speaks for itself. The results are obvious to anyone who cares to observe them.

Physically, I get very tired from real work that uses muscles and burns calories and accomplishes things that matter. But it's satisfying exhaustion. It comes from productive effort, not from the mental exhaustion that comes from sitting still while your brain gets tortured by meaningless activities. Activities designed to occupy time without producing anything of value, except paychecks for people who attend meetings about meetings. Where nothing gets decided, and everything gets postponed until the next meeting, where the same nothing will be discussed again.

I fall asleep in seconds as soon as my head hits the pillow, because my body is tired from work that matters. My mind is at peace because I spent the day doing things that align with my values. Not things that contradict everything I believe about how life should be lived, and what work should accomplish for the people who do it, rather than just for the people who profit from it.

Before, I would stay up all night rehearsing emails and conversations. Replaying interactions with coworkers and supervisors who treated me like a problem to be managed rather than a human being. Someone with knowledge and skills that could be used to solve actual problems, if anyone cared about solving problems, instead of just managing people and processes that exist primarily to justify management positions. Positions that serve no useful purpose except creating hierarchy and control structures.

The Corporate Sleep Theft

Corporate life stole my sleep. Through stress and anxiety about workplace politics, real-life politics, performance evaluations, and project deadlines that served no useful purpose except keeping people busy and worried. Job security could be threatened by arbitrary decisions made by people who knew nothing about the actual work being done. That work, done by employees, generated the profits that paid for management salaries and corporate overhead.

The mental rehearsal of workplace interactions was its own kind of torture. It extended work stress into my personal time. Time that was supposed to offer rest and recovery from the daily grind. A grind that required pretending meaningless work mattered, solely because it generated revenue for people who treated employees like expendable resources. Once those employees became too expensive or too difficult to manage, they were replaced. Corporate policies are designed to maximize compliance and minimize independent thinking.

Sleep became my escape from consciousness, not natural rest. Because consciousness during corporate employment meant full awareness of how much life was being wasted on tasks that served other people's interests. Those tasks offered minimal compensation and no meaningful satisfaction. And all the while, people know they are capable of more fulfilling and productive work, if only they could escape the systems trapping them in jobs that slowly killed their souls under the promise of security that never materialized.

Now my sleep is natural. It follows real exhaustion from work that produces results I can see, touch, eat, sell, and take pride in. Work accomplished on my terms, by my standards. Not by the metrics of people who measure productivity in ways that ignore actual value, human satisfaction, or meaningful contribution to survival, wellbeing, or community prosperity.

The Death of Weekends and the Birth of Real Living

Weekends and holidays lose their meaning when you spend your days with animals, flowers, and gardens. Doing work that feeds your body and soul. Corporate work pretends to offer security, but it kills both body and soul slowly while barely covering the cost of living in systems designed to extract as much value as possible for the lowest compensation.

Who needs weekends when every day is spent doing work you love, in places you want to be, with creatures that depend on you and appreciate your care? They don't play political games. They don't create artificial drama to climb a corporate ladder at your expense.

Relief from work stress becomes irrelevant when your work no longer creates stress. When it serves your purposes, not someone else's profit margin.

Time off during holidays means nothing when you never have to go back to a job you hate. Because you've built work that aligns with your values, serves your needs, and brings real meaning. Every day feels like a holiday compared to the soul-crushing routine of corporate life, where people are treated like resources and managed like inventory.

The line between work and personal time vanishes when your work is a reflection of your values instead of a compromise. When your work builds independence, not dependence. When it supports your goals instead of undermining them.

Natural rhythms replace artificial schedules. Your days follow the sun, not a calendar full of arbitrary time blocks. You stop fragmenting your life into meetings and metrics designed without any regard for human energy cycles, seasons, or real biological needs. But acknowledging those needs would interfere with standardized productivity, which is why corporate systems ignore them.

The Fake Urgency Industrial Complex

Corporate urgency is manufactured drama. It keeps people stressed, reactive, and compliant. It serves the convenience of management, not operational necessity. Most deadlines exist to apply pressure, not to meet actual constraints, customer needs, or logical timelines.

Real urgency is different. It happens when animals get hurt, storms threaten crops, or equipment breaks during critical tasks. These are genuine problems with visible consequences and obvious solutions. Problems solved through direct action, not meetings and management approval.

Most corporate emergencies are artificial. They're about fixing problems caused by previous corporate decisions. Preparing reports for meetings where nothing gets decided. Responding to uninformed requests made by people who don't understand the work or the time it requires. The result? Corners get cut, bigger problems emerge later, and the cycle continues.

Farm emergencies are real. Animals die if you don't act. Crops fail. Equipment breaks in ways that stop everything until it's fixed. These moments demand knowledge, tools, and effort. Not bureaucratic delays.

Solving real problems with real skills builds true confidence. It strengthens competence that carries over into life. That's the opposite of corporate problem-solving, which is mostly about navigating politics and processes that have nothing to do with actual effectiveness.

When you face real problems with real consequences, you grow. You become capable. That self-assurance helps you handle anything. Meanwhile, corporate work just creates anxiety. Because the problems aren't real, the solutions don't matter, and the skills don't transfer to anything that helps you outside that world.

The Revolution of Not Checking Your Phone

The longest I've gone without checking my phone is several hours. Why? Because I was immersed in work that demanded full attention. Physical work that satisfies in ways digital distractions never can. Most people today can't go several minutes without checking their phones.

Phone addiction grows in jobs that lack meaning. People reach for stimulation when they're bored or miserable. They do tasks they don't care about for people who see them as replaceable labor, nothing more.

Animals don't care about your schedule or your phone. They don't need you to be digitally connected. They need food, care, and presence. Giving them that pulls you out of systems built to sell you dog shit you don't need. Systems that distract you from what actually matters.

When your work is physical, and when lives depend on you, you don't want distractions. They interrupt you. They don't help. You stop needing stimulation because your work already gives you what you used to look for in your phone: purpose, satisfaction, challenge.

Not checking your phone isn't a digital detox. It isn't about discipline. It's the natural result of having real work that captures your full attention and fills your day with something worth being present for. When life itself is satisfying, you don't need an escape.

Chapter 17: The Entitlement of the Trapped

We don't deal with people's reactions to our lifestyle choices because we don't give a shit what anyone thinks about decisions that serve our interests. Their comfort with systems that are slowly killing them isn't our concern. They pretend everything is fine. They call it normal. They call it sustainable. Even when it's barely survivable for anyone trying to live instead of just comply until death feels like relief.

This is our farm and our home. If people don't like how we live, what we produce, or how we operate, based on our values rather than their expectations, they can go fuck themselves. They can find someone else to judge, criticize, and offer unsolicited advice to. We've chosen to live outside systems built to keep people dependent, compliant, and afraid of alternatives that might actually work if they were willing to do the work to build them.

Some people think they're entitled to our food just because they have money. But money alone doesn't entitle anyone to what we produce through labor, resources, knowledge, time, and care for animals that rely on us for everything. Those animals live full, healthy lives that end humanely for nutritional purposes, not to satisfy some industrial model obsessed with efficiency over welfare.

That food is worth more in our own bellies and in our freezer than it is in the hands of people who don't value the work behind clean meat and fresh eggs. We operate on standards that prioritize animal welfare and quality over cost-cutting and profit. Industrial systems treat animals like inputs and customers like revenue streams. Not people.

When customers act entitled, we remind them who controls the food. Their money means nothing compared to our need for nourishment and food security. That can't be bought when supply chains fail, prices spike, or quality drops so low that store-bought food becomes unsafe or inedible for anyone who knows what real food tastes like.

The Mercedes Lady and Her Country Girl Bullshit

There's a certain type of person who performs identity like they're auditioning for a role they'll never get. You can spot them from a mile away. All the right props, all the right words, none of the actual understanding that comes from living the thing they're pretending to be.

There was a lady, driving a Mercedes with her husband, who visited our farm stand. She claimed to be a "country girl," which was clearly bullshit from the moment the words left her mouth. Our chicken eggs go quick, especially during a nation-wide egg shortage, and we were sold out during their visit. When they asked if we had any chicken eggs, I offered them free quail eggs to try instead. With disgust on her face, she turned down free quail eggs, saying "I'm a country girl, but not that country."

Real country people don't turn down free eggs from neighbors trying to build community and share their harvest. Especially not when those eggs come from a well-run farm with more food than one family can eat.

This is the difference between rural theater and rural reality. Rural theater is driving your luxury car to the cute farm stand to buy authenticity by the dozen. Rural reality is understanding that today's abundance could be tomorrow's scarcity. That community gets built through mutual aid, not mutual performance.

She threw out the "country girl" line and then trailed off saying "...but not that country" like we were hillbillies or beneath her. Like our quail eggs, fresher and more nutritious than anything in a store, weren't good enough for even someone who grew up "in the country." As if they weren't better than factory eggs laid by birds stuck in cages, fed garbage, and never seeing sunlight. As if quail eggs aren't treated as a delicacy in high-end restaurants.

The next time she showed up for chicken eggs, I asked if she wanted quail again, for free. When she said no, I told her if she was too good for our free quail eggs, then she didn't need our chicken eggs either. She could get the fuck off our farm and buy from wherever "country girls" shop when they're too sophisticated to accept gifts from actual farmers.

The beautiful thing about our position is that we can afford to tell customers like her to fuck off. Most small business owners dream of having that luxury but can't risk it because they need every sale to stay afloat. We don't. Our business plan doesn't depend on the public's approval or their sporadic purchasing decisions. We're like the parent who stays home and provides childcare instead of using their income to pay for daycare. We subsidize our own food costs by producing what we eat. If nothing sells, we still eat better than anyone shopping at the grocery store.

Our place of business is also our home, and it's paid off. No rent, no mortgage payments that depend on customer traffic. When people don't buy our eggs, we eat them ourselves. When they don't buy our meat, it goes in our freezer. We're not sitting on inventory we can't use. We're sitting on food that's more valuable to us than whatever price we could get for it.

This setup gives us leverage that traditional businesses can never have. We don't worship customers because we don't need them to survive. They should be grateful for access to food this clean, this fresh, this ethically produced. But if they're not, that's their loss, not ours. We know what we have. We know what it's worth. And we're not going to devalue it to accommodate people who mistake convenience for entitlement. Further, we aren't going to sell it to people who don't value it and take that opportunity away from someone who would.

The financial cushion we built ensures we never have to compromise our standards for a sale. We didn't spend everything on the land purchase. We kept enough to guarantee our independence from fickle customers and seasonal fluctuations. Because customers never bail you out when you need it. That would require them to have money at the exact moment you need help, and it would require them to give a fuck about you personally. Neither of those things are reliable business strategies.

People who reject free food while claiming rural credibility are performing an identity they never earned. What they really want is a curated farm experience. Something that makes them feel connected to "authenticity" without actually building relationships with farmers or understanding food production. They want the aesthetic without the reality. The credibility without the work. The identity without the values.

This woman wanted rural identity without rural relationships. She wanted to feel country without being a part of our community. And that's exactly how most people approach everything that matters. They want the performance without the practice. The aesthetic without the ethics. The identity without the reality.

We've created something more powerful than a business. We've created leverage. Food is the ultimate currency because everyone needs it, and quality food commands respect. When you control food production, you control something that can't be outsourced, can't be digitized, and can't be made irrelevant by technological disruption. People might not need another app or gadget, but they'll always need to eat.

The strategy isn't complicated. We're selling nostalgia to an aging population who remembers when food actually tasted like something. They become our sales force, teaching their kids and grandkids what real meat tastes like, what fresh eggs should look like. Rabbit and quail aren't just products. They're connections to a food culture that industrial agriculture nearly destroyed. The first sale educates the whole family, and suddenly we have customers across generations without spending a dime on marketing.

When people taste our rabbit prepared the way their grandparents used to make it, they're not just buying meat. They're buying back a piece of their heritage that they didn't realize they'd lost. When their kids taste quail for the first time, they understand why their great-grandparents considered it a delicacy worth the effort to hunt and prepare.

This isn't manipulation. It's honesty about what we're selling and why it matters. We're not hiding anything or making false promises. We're just producing something so much better than the industrial alternative that the product sells itself once people try it. Everything downstream takes care of itself because the food does what real food is supposed to do. It makes people remember what they've been missing.

The Cute Farm Stand Delusion

Most people don't understand that our food is about survival, not entertainment. Not aesthetics. Not teaching suburban kids a sanitized version of farm life. They want animals, not animal husbandry. Gardens, not food systems. They want an experience without learning what it takes to feed a family without relying on a fragile, outsourced system.

They see our farm as cute, not essential. Decorative, not functional. A hobby, not a business. But this farm is our job. Our income. Our survival. We stepped away from employment systems that treat workers like tools, and from an economy that treats consumers as targets meant to keep spending, never producing.

Farm stands draw people who want to feel like they're supporting local agriculture. But most don't care about what actually makes it viable. Things like paying for veterinary care, feed, equipment, utilities, and taxes. You can't barter with farm products to pay the electric company or the IRS.

The cute farm aesthetic appeals to people who romanticize rural life but don't understand the risk, the effort, or the investment involved.

Real food production isn't a photo op. It's hard. It's constant. It's full of uncertainty. And it's not financially sustainable without customers who are willing to pay prices that match the reality.

Tourists want Instagram stories and feel-good narratives about sustainability. But they flinch the second you show them the price tag that comes with feeding animals clean food, raising them humanely, and not cutting corners. They want the fantasy without funding the truth. And the truth is, our prices are already lower than what they'd pay for factory meat disguised as premium. Meat that comes from animals raised in misery, processed in filth, and sold in packaging that costs more than the protein inside.

The "Doing You a Favor" Mentality

Some customers act like they're doing us a favor by shopping here. As if buying our products is a form of charity instead of a transaction that keeps our business alive. We use our time, land, knowledge, and care to raise animals who live good lives until they become food. Those eggs and that meat exist because of daily work they'll never see.

This "favor" mindset shows a complete misunderstanding. We aren't a non-profit petting zoo. We aren't here to entertain you or teach your kids a lesson in gratitude. We're here to make a living. And that means we have to charge what the food is worth because it costs us to produce it.

We don't need charity. We need customers who understand what they're buying. Customers who know that our food costs more than the industrial equivalent because it's better. Because the methods behind it are better. Because they align with values most people say they care about until it comes time to pay for them.

The right customers don't flinch. They buy regularly. They tell others. They become part of our community. And they understand that you can't support small farms by demanding factory prices. Those prices exist because of externalized costs like animal cruelty, environmental damage, and labor abuse. We don't operate that way. When people expect small farmers to match industrial prices while maintaining higher standards, they reveal their ignorance. They expect us to subsidize their food with our labor, our health, and our bank accounts. That's not sustainable. And we won't do it.

But our prices still beat grocery stores and farmers markets every time. Our rabbit meat is cheaper than the gray slop Walmart calls beef. Our eggs are cheaper than the so-called "farm fresh brown" ones sitting under fluorescent lights, already refrigerated and already past their prime. Our whole quail? Only seven dollars per pair for a fucking delicacy. All hatched, raised, and processed by us. Find us a better price on quail, or tell us you think that's expensive.

What people pay top dollar for in high-end restaurants; we raise and eat for pennies on the dollar. Ever see how much a rabbit leg costs at Le Lapin Sauté in Canada? The amount you'd pay there for a single leg would feed a lot of rabbits each month.

Chapter 18: Real Security vs. Paper Promises

The Illusion of Retirement Security

I realized 401(k) and TSP accounts were fake wealth the moment I got penalized for accessing my own money. This was money I earned. Money taken from my paycheck through payroll deductions that reduced my take-home pay in exchange for vague promises of future security.

That security was conditional. It depended on market performance, on systems controlled by people who profit whether your investments go up or down. They get paid either way. Your future hinges on returns that may never come, right when you're too old to work and need income most.

Early withdrawal penalties expose the truth. These accounts weren't designed for your protection. They were designed to trap your money. Real wealth doesn't require permission slips. Real security isn't held hostage by penalty clauses that punish you for needing what you earned. These institutions thrive by locking you out of the very funds they claim are yours.

If employer contributions are part of your compensation, why do you have to wait decades to access them? You're working now. They're profiting now. But if you leave, that money is suddenly out of reach or worth less than it was promised to be.

The whole system is built to keep workers in line. Retirement benefits aren't rewards. They're golden handcuffs. They discourage you from quitting. They tie you to jobs that don't care if you're miserable, as long as you're profitable.

And all of it rests on the blind assumption that markets will keep climbing for decades. As if economies never crash. As if political and financial systems don't break. As if the future will behave like the past. When the truth is, the collapse might come long before your savings ever matter.

The Mirage of Financial Safety

Financial safety, when it's built on paper assets you don't control, is a lie. It's a comfortable story told to keep you calm. A myth sold by people who profit from your belief in it.

You're told to trust institutions. To hand them your money. To let them manage it for a fee. Whether they help you or screw you over, they still get paid. Their systems are designed to confuse you because complexity keeps them profitable.

But real security doesn't live in spreadsheets. It lives in what you can touch. What you can grow. What you can fix. It lives in skills, infrastructure, and self-sufficiency.

You want food security? Grow your own. Don't depend on someone else to feed you. Especially not someone making it in ways you don't understand, with ingredients you've never seen, shipped through fragile supply chains that collapse the moment things get difficult.

You want energy security? Generate your own power. Don't rely on utility companies that raise your rates every year and threaten to shut you off if you fall behind. Or when their grid fails, gets hacked, or breaks down from neglect.

Housing security doesn't come from a 30-year mortgage. It comes from owning the roof over your head. If you don't, all it takes is one missed paycheck or one economic downturn and you're out.

Water security is no different. Find your own source. Don't rely on city systems that can be contaminated, mismanaged, or broken for weeks while you're left boiling water and hoping someone competent finally shows up.

The TSP Realization That Broke the Spell

When I cashed out my TSP early, it wasn't reckless. It was freedom.

The penalties didn't scare me. They confirmed what I already knew. The account wasn't there to help me. It was there to manage me. To keep me in line. To hold my money hostage until someone else decided I had waited long enough.

Those early withdrawal fees are nothing more than punishment for choosing freedom. They penalize you for needing what you earned. They expose the truth about the system. It's not built for your safety. It's built to keep you obedient.

After taxes and penalties, I took what was left and we used it. Right away. We built real things. We bought land. We set up infrastructure. We invested in tools that give us independence. Not promises. Not forecasts. Not some padded retirement account with rules made by people who don't live like us.

That money, even reduced by fees, did more for us than a lifetime of theoretical growth ever would. Because it gave us control while we still had energy to use it. It gave us something real. Not a vague future. Not a number on a screen that disappears when markets crash or policies change.

And no, it wasn't irresponsible. What's irresponsible is trusting a system designed to profit off your blind compliance. What's reckless is giving away your future to people who benefit from your silence.

The Defense of a Broken System

People defend 401(k)s like they're sacred. They repeat talking points about compound interest and tax advantages like scripture.

But those so-called advantages come with assumptions. They assume the market will behave. They assume political stability. They assume your currency will hold value for the next forty years. None of that is guaranteed.

What you're really being told is this: Trust a system that can't promise anything, and wait four decades to see if it worked.

What Happens When You Say This Out Loud

When you question retirement dogma, people get uncomfortable. They've built their futures on these plans. They've never challenged them because everyone else accepts them. Challenging them means facing the truth: they were sold a trap disguised as a gift.

They bought financial products wrapped in the language of security. But those products were built to serve the people selling them, not the ones buying in.

The complexity isn't accidental. It's the entire point. It hides the fees. It hides the risk. It keeps you confused, dependent, and easy to manage.

Employer Matching: The Scam Explained

Tell someone that employer matching is just delayed compensation they can't actually access, and you'll watch the lightbulb flicker.

You do the work now. They get the value now. What you get is a promise. A promise you can't access unless you stay put, behave, and hope the market plays nice.

Employers get tax breaks the second they "contribute." You might get one too, but only if you don't need the money when you actually need it. Try to use it early, and you're penalized for taking care of yourself.

By the time you're allowed to touch it, it's too late to enjoy it. Your body is worn out. Your best years are behind you. And what's left of that money might cover survival, but it won't buy back time.

Real Wealth Doesn't Require Permission

Real wealth isn't an account balance. It's not a graph. It's not a password-protected login screen.

It's what you can use, touch, and control.

It's your freezer full of meat. Your solar panels. Your water catchment system. Your root cellar. Your workshop. Your knowledge. Your time.

It's growing your own food instead of buying "certified organic" from a store that sources it from 1,000 miles away.

It's vegetables you can eat tonight, not bonds that might pay next quarter if the borrower doesn't default.

It's your ability to fix your roof, wire your shed, sharpen your blades. Not hiring someone else for tasks you were convinced you couldn't do.

Part IV: Using the System to Leave the System

Chapter 19: Weaponize Social Media

Before we opened our doors, we gave ourselves 120 days to build our email list. That was the plan. Use social media with intention, not addiction. Build our list. Then leave. We used Facebook, Instagram, and TikTok because they're the easiest platforms to push content to for us. Notice I said used. As of the publication of this book, we have dropped all social media and rely entirely on our newsletter. It is direct, honest, and ours, free from algorithms, ads, and platforms that profit from distraction. We reclaimed our time by using these platforms strategically rather than getting sucked into endless scrolling and engagement designed to harvest our attention and sell it to the highest bidder. Now we don't use these platforms at all.

The approach was simple. We completed our posts and closed the app. We didn't stick around to watch metrics that didn't matter, respond to every comment from people who didn't understand what we're trying to do, or get caught up in the algorithm's attempt to keep us scrolling until our eyes bled and our minds turned to mush.

We used the platforms as tools, not entertainment or validation systems. These platforms are designed to be addictive, to keep you engaged long enough to show you advertisements and collect data about your behavior that can be sold to people who want to manipulate your decisions and extract money from your weaknesses.

The Algorithm as Oppressor

We know how the algorithm works, and we don't care about making it happy because algorithms serve the people who own them, not the people who feed them content. If people want to find us, they will.

We're not optimizing for reach or engagement or viral moments that turn human experiences into entertainment for people who are slowly dying of boredom. We do not need to. Remember, we have no mortgage, and we have broken free from nearly every system architected to pin us down.

The algorithm is designed to show you what keeps you scrolling, not what helps you live better. It promotes conflict over cooperation, outrage over understanding, and consumption over creation, because those emotions drive engagement and engagement drives profit for platforms that profit from human misery.

We're not trying to convert anyone or build a movement or become influencers selling dreams to people who are trapped in nightmares. The algorithm can promote us or bury us. It does not change what we're doing or why we're doing it because our purpose does not depend on permission from machines designed to serve masters we will never meet.

Content as Resistance

Our content strategy was straightforward: show the results, not the process. We worked for these results. The freedom, the independence, the peace that comes from not participating in systems designed to consume you. That is what we documented. Not every struggle or setback or moment of doubt. Struggle porn serves the system by making people think that suffering is noble instead of unnecessary.

We didn't care who responded to our content. Caring about other people's responses is a trap that keeps you performing for an audience instead of living for yourself. Some people relate to what we're sharing because they're ready to hear it. Some people think we're crazy because they cannot imagine life outside the cage they have learned to call home. Some people ignore us completely because they are too busy dying slowly to pay attention to people who chose to live.

All of those responses are fine. We're not responsible for other people's choices, reactions, or their ability to understand what is possible when you stop believing what you're told about what is impossible. We're not trying to save anyone. You cannot save people who do not want to be saved. Most people prefer familiar suffering to unfamiliar freedom.

Platform Independence as Survival Strategy

Platform independence is crucial. In five months, any of these platforms may not be around, or they may change their rules in ways that make our content invisible, or they may decide that our message threatens their profit model and remove us entirely.

The rules change constantly. Accounts get banned or shadowbanned for reasons that make no sense except as exercises in arbitrary power. Revenue sharing gets cut or eliminated when platforms decide they have given creators too much of the money that the creators generated. Ad rates fluctuate based on algorithms designed to maximize platform profit while minimizing creator compensation.

We don't depend on any platform for our income or our identity because dependence on systems you don't control is just another form of slavery with better marketing. We treat social media like we treat everything else: useful when it serves our purposes, irrelevant when it doesn't. We're not influencers trying to build personal brands that can be monetized by people who see human attention as a commodity to be harvested and sold.

Authenticity as Weapon

We're not entrepreneurs trying to scale businesses that would require us to hire employees and manage people and create the same kind of extractive relationships we escaped from. We're just people sharing what we've learned while building the life we wanted, people who refuse to pretend that normal is healthy or that the system serves anyone except the people who own it.

The content is honest, not polished. Real moments, not staged performances for people who mistake entertainment for education. Actual information, not clickbait or outrage farming designed to generate engagement from people who confuse anger with action.

We share what works, what doesn't work, and what we're still figuring out. No fake expertise or false promises. No guru nonsense designed to sell courses to people who are looking for someone else to solve problems they need to solve themselves.

We didn't chase trends or try to game the system. We didn't post at optimal times because optimal times are determined by algorithms designed to serve platform profits, not creator success. We didn't use trending hashtags that have nothing to do with our content because using irrelevant hashtags is lying to attract people who aren't looking for what you're actually offering.

We didn't collaborate with other creators because most creators are selling something we don't believe in, promoting lifestyles we escaped from, or building audiences they plan to exploit for profit. We shared when we had something worth sharing and ignored the platform when we didn't.

Building Real Connection in a Fake World

The goal isn't to build an audience. The goal is to connect with the people who are ready to hear what we have to say, people who are already questioning the systems we escaped from and looking for proof that alternatives exist.

Quality over quantity. Authenticity over algorithm optimization. Real connection over engagement metrics. Social media can be a tool for sharing your message without letting it control your life, but only if you approach it with clear boundaries and realistic expectations about what it can and can't do for you. Most people use social media to avoid dealing with problems that social media makes worse while providing temporary relief from symptoms it helps create.

Use it, don't let it use you. Speak plainly. Connect with people who matter, don't chase people who don't. Build something real, don't optimize for metrics that don't improve your actual life.

The platforms want you to believe that success means having millions of followers and viral content and brand partnerships with companies that profit from human suffering. Real success means building a life you don't need to escape from, sharing that life with people who understand why escape was necessary, and ignoring the noise from people who mistake captivity for security.

Chapter 20: Digital Products, Real-World Payoff

We sell what we actually make and what we actually know. That's it. No bullshit courses about "mindset" or "crushing it." No affiliate links to products we've never used. No fake guru nonsense about scaling your way to millions while working four hours a week.

Our business is simple because our life is simple: rabbit manure for gardens, hatching eggs for people starting flocks, meat from animals we raised and processed ourselves, flowers we picked by hand. Plus information products for people who want to learn what we learned the hard way.

The difference between us and every other "farm business" you see online? We actually do this. We're not lifestyle bloggers cosplaying as farmers. We're not selling you a dream we've never lived. When I tell you how to process a rabbit, it's because I processed one this morning. When Alexys explains companion planting, it's because she's got dirt under her fingernails from doing it yesterday.

The Real Products That Actually Matter

People buy our rabbit manure year-round because gardens need fertilizer and our rabbits shit consistently. It's not seasonal. It's not trendy. It's triple-sifted, premium fertilizer that makes plants grow better than the chemical garbage from the store. We know because we use it ourselves.

Our hatching eggs go to people who understand that raising your own food starts with raising your own breeding stock. Quail eggs for people who want the best protein source per square foot. Chicken eggs for people who want variety and color and actual nutrition instead of those pale, flavorless things from factory farms.

The meat we sell comes from animals that lived good lives and died ethically. No factory farm torture. No antibiotics. No hormones. No wondering what they ate or how they were treated. You know exactly where your protein came from because you bought it from the people who raised it.

Our flowers are cut fresh and hand-tied by Alexys because store-bought flowers are shipped from thousands of miles away, pumped full of preservatives, and cost three times what ours do. Ours last longer, smell better, and don't support an industry that treats beauty like a manufacturing process.

Information Products That Don't Suck

The digital stuff we create isn't about "building your personal brand" or "leveraging your expertise." It's about sharing what works and what doesn't work for people who are serious about changing their lives instead of just talking about it.

We make courses about processing animals because most people have been convinced they can't do it themselves. They can. It's not complicated. It's just been hidden from them by systems that profit from their helplessness.

We write about quail because quail are efficient as hell and most people have never considered them. Six weeks from hatch to harvest. Six weeks to laying eggs. Try finding that efficiency in any other protein source.

We consult with people who are planning their own exits because we remember what it felt like to be trapped and looking for someone who had actually done what we were trying to do. Not someone who talked about it. Someone who lived it.

Why We Don't Scale Like Every Guru Tells You To

Every business expert would tell us we're doing it wrong. We should hire employees. We should automate everything. We should create passive income streams and build systems that run without us.

Fuck that. We escaped one system designed to consume our lives. Why would we build another one?

The goal isn't to get rich. The goal is to stay free. Rich people have problems we don't want. Rich people have employees who depend on them, investors who own pieces of their businesses, and growth targets that require them to compromise their values for profit margins.

We control every aspect of our business because we control every aspect of our lives. Our products are priced fairly because we don't have shareholders demanding maximum extraction from every transaction. We can afford to be honest because we don't depend on hype and manipulation to generate revenue.

When customers become entitled assholes, we fire them. When people try to negotiate our prices, we tell them to shop elsewhere. When someone wants us to change how we operate to accommodate their preferences, we remind them that our farm runs according to our values, not their convenience.

That kind of freedom only exists when you own your life and your business completely. No debt. No investors. No business partners. No employees whose salaries depend on your ability to generate revenue, whether you feel like working or not.

The Platform Independence Strategy That Actually Works

We used social media platforms because they were useful tools for reaching people, but we don't depend on them for our income or our identity. Instagram could have banned us, but we would still eat dinner. Facebook could have changed their algorithm and we'd still pay our bills.

The real business happens through our website, our email list, and direct relationships with customers who know where to find us when they want what we produce. We're not influencers trying to build audiences we can monetize through advertising revenue that fluctuates based on engagement metrics we can't control.

Platform independence means treating social media like a hammer: useful when you need it, irrelevant when you don't. We don't live on these platforms. We don't refresh our notifications every five minutes. We don't optimize our content for algorithm approval.

We post when we have something worth sharing. We respond when people ask genuine questions. We ignore the noise from people who confuse online drama with real interaction. The platforms serve us, not the other way around.

Money Comes From Value, Not Hype

Our products sell because they're better than the alternatives, not because we're good at marketing. Our rabbit meat tastes better than store-bought garbage. Our eggs are fresher. Our flower arrangements are hand-tied by Alexys using what's blooming in-season here in West Virginia, without high tunnels or climate control. No two bouquets are the same.

We don't need to convince people to buy what we're selling. The products convince people. The results convince people. The difference in quality is obvious to anyone who tries what we produce compared to what they're used to buying from industrial sources.

When you make things that are genuinely better, marketing becomes education instead of manipulation. You explain what you do differently and why it matters. You let people decide if they value what you're offering enough to pay what it costs to produce it properly.

Most businesses spend more money on advertising than we spend on our entire operation because most businesses are selling products that aren't actually better than cheaper alternatives. They need to create artificial demand through psychological manipulation and artificial scarcity.

We create real demand by solving real problems with real solutions. People need food. They need clean food. They need food that doesn't come from industrial systems that treat animals like manufacturing inputs and customers like suckers. We provide that.

The Business Model That Serves Life Instead of Growth

Our business model isn't scalable in the traditional sense, and that's exactly why it works for us. We can't franchise rabbit processing or outsource flower arrangement or hire employees to take consultation calls about lifestyle changes they've never experienced.

The business serves our independence instead of requiring us to sacrifice our independence for business growth. We work the hours we choose. We serve the customers we want to serve. We make enough money to cover our expenses and save for the future without compromising our values or burning ourselves out chasing metrics that don't improve our actual lives.

Success means sustainability, not expansion. It means maintaining what we've built without killing ourselves to build more. It means having the luxury of saying no to opportunities that don't align with our goals, even when those opportunities look good to other people who mistake busy for productive.

We're not trying to build an empire. We're trying to build a life. The business supports our life, not the other way around. When business decisions conflict with life decisions, our life wins every time.

That's the difference between building something that serves you and building something that consumes you. We've been consumed before. We're not going back.

Chapter 21: Email as Homestead

Email cuts through the bullshit. No algorithm decides who sees what we send. No platform can make our messages disappear because they don't align with advertising revenue goals. No corporation controls our ability to communicate with people who actually want to hear from us. This is only true because we can bring our email list wherever we want.

When we send an email to our subscribers, it lands in inboxes. When people subscribe, they get what they signed up for. When we have something important to share, we share it directly without hoping the social media gods will allow our content to reach the people who chose to follow us.

Email lists are the last digital communication channel that belongs to the sender instead of the platform. Everything else is rented space in someone else's empire. Email lists are owned land in the digital world.

Social media followers aren't yours. They're numbers on platforms that can disappear with policy changes, account suspensions, or algorithm updates that make your content invisible to people who explicitly chose to see it. Platforms that profit from your content while limiting your reach unless you pay for the privilege of communicating with your own audience.

An email list is different. Those addresses belong to you. The relationships belong to you. The communication channel belongs to you. No middleman collecting data about your subscribers to sell to advertisers. No artificial intelligence deciding who deserves to see your messages based on engagement metrics that change whenever platforms need to squeeze more advertising revenue from their users.

We treat our email list like digital homestead land. Something we own, cultivate, and protect. Something that provides security and independence from systems designed to extract value from our work while providing minimal compensation and no guarantees.

Building an email list is like clearing land, preparing soil, and planting crops that will feed you for years instead of sharecropping on someone else's platform where they control the harvest and you get whatever scraps they decide to leave you.

Real Communication vs. Performance Art

Email lets us have actual conversations instead of performing for audiences who might not even see what we post. No character limits that force complex ideas into oversimplified sound bites. No comments sections filled with people who confuse argument with engagement.

When we send newsletters, we're talking to people who asked to hear from us about topics they care about. People who value what we're building and want updates about things that might matter for their own lives and decisions.

We don't send daily emails because we don't have daily insights worth sharing. We don't follow posting schedules that require us to generate content whether we have something valuable to say or not. We communicate when we have information, updates, or perspectives that serve the people who subscribed to receive them.

The relationship is voluntary on both sides. People can unsubscribe anytime they want. We can choose what to share and when to share it. No platform policies dictating acceptable content. No artificial intelligence monitoring our language for advertiser-friendly compliance.

Direct Sales Without Soul-Selling

Email marketing doesn't have to be the sleazy manipulation most people associate with the phrase. It can be honest communication about products and services that solve real problems for people who actually have those problems.

When we have rabbit meat available, we let our email subscribers know first. When we're taking orders for seasonal flowers, email gets priority access. When we're offering consultation calls, the people on our list hear about it before anyone else.

This isn't bait-and-switch manipulation or artificial scarcity designed to pressure people into buying things they don't need. It's practical communication about genuine availability of products and services that our subscribers have expressed interest in receiving.

We don't use psychological triggers or sales funnel tactics that treat potential customers like marks to be conquered. We share information about what's available, why it might be useful, and what it costs. People decide for themselves if they want what we're offering.

The goal is building long-term relationships with people who value what we produce, not extracting maximum short-term revenue from people we'll never interact with again. Customers who trust us buy from us repeatedly and refer people they care about. That's more valuable than one-time sales to people who feel manipulated into purchasing.

Platform Independence as Insurance Policy

Email lists survive platform changes, policy updates, and corporate decisions that can destroy years of audience-building work overnight. Platforms that seemed permanent have disappeared. Features that creators depended on have been eliminated. Algorithms that once promoted content now bury it unless creators pay for visibility.

Our email list protects us from the whims of people who profit from our content while controlling our access to the audiences we built through our own work. When platforms change rules, our communication channel stays the same. When algorithms shift, our messages still reach inboxes.

Platform independence through email means we're not vulnerable to decisions made by people who see creators as free content generators whose value can be extracted through advertising revenue while providing minimal compensation and no long-term security.

We use social media as recruitment tools for our email list, not as primary communication channels. Social media introduces people to what we're doing. Email builds relationships with people who want to stay connected.

Building Community vs. Collecting Data

The people on our email list aren't numbers to be optimized for conversion rates. They're people who share interests and values related to independence, self-sufficiency, and alternatives to systems that extract more value than they provide, and people who like good, clean food too.

We don't segment our list based on buying behavior or demographic data. We don't A/B test subject lines to maximize open rates. We don't track click-through percentages or analyze engagement metrics that treat human attention like a resource to be harvested and monetized.

The emails we send provide value whether people buy anything or not. Updates about what we're learning. Insights about challenges we're facing. Information about techniques that work and approaches that don't. Stories about daily life that illustrate larger principles about independence and intentional living.

Building community through email means respecting people's time and attention instead of exploiting it. Sharing useful information instead of creating artificial urgency. Treating subscribers like allies instead of targets in a sales process designed to overcome objections and extract revenue.

The Newsletter as Documentation

Our email archive documents our journey from corporate employment to farm-based independence. The progression of learning, building, succeeding, and failing that can't be captured in social media posts designed for immediate consumption and algorithmic distribution.

Email newsletters create permanent records of thoughts, decisions, and experiences that provide context for people who want to understand not just what we did, but how we thought about problems and opportunities as they arose.

This documentation serves people who are planning their own transitions and want to learn from our experience without having to repeat our mistakes. It also serves us as a reference for decisions we made and approaches we tried when memory fades and details become unclear.

The archive becomes a resource that grows more valuable over time instead of disappearing into social media timelines that bury old content under new posts that demand immediate attention but provide no lasting value.

Email as Digital Sovereignty

Controlling your communication channel is a form of digital sovereignty that becomes more important as platforms consolidate power over information distribution. When a few companies control how billions of people access information and communicate with each other, those companies control public discourse and economic opportunity.

Email represents resistance to that consolidation. It's a distributed system that can't be controlled by any single entity. It works without permission from platform owners. It survives corporate policy changes and algorithmic modifications designed to extract more profit from user attention.

Building and maintaining an email list is an act of digital independence that parallels physical independence through food production, energy generation, and financial sovereignty. It's about controlling essential infrastructure instead of depending on systems owned by people whose interests don't align with yours.

We treat our email list like we treat our land, as something that provides security, independence, and the ability to communicate and create value without asking permission from people who profit from our dependence on their systems.

Email is our digital homestead. We cultivate it, protect it, and use it to build relationships and share value with people who understand why independence matters more than convenience, why ownership matters more than access, and why direct communication matters more than algorithmic amplification.

The list belongs to us. The relationships belong to us. The communication channel belongs to us. That's digital sovereignty in a world designed to make you a sharecropper on someone else's land.

Chapter 22: The Money Was Always Yours

People love to tell us that cashing out retirement accounts is "irresponsible." That we "threw away our future" by taking early withdrawal penalties. That we should have been "patient" and waited until we were 65 to access money we earned in our 20s and 30s.

But here's what they don't understand: that money was already ours. Every dollar in those accounts came from our paychecks, our labor, our time. The only thing standing between us and our own money was a system designed to trap it there until we were too old to use it for anything meaningful.

The whole retirement account scam works because they convinced us that accessing our own earned money is gambling, while trusting institutions to manage our future for 30 to 40 years is somehow "safe." But let's be honest about who's really gambling here.

The Compensation Scam in Detail

Let's break down exactly how the retirement account shell game works, because understanding the mechanics makes it easier to see why accessing "your own money" isn't irresponsible; it's overdue.

When you get a job offer for $75,000 plus "up to 6% matching," what they're really saying is: "We could pay you $79,500 in actual salary, but instead we'll pay you $75,000 cash and put $4,500 in an account you can't touch for 30 years. Oh, and you have to contribute $4,500 of your own money to get our $4,500, so you're actually getting $70,500 in take-home pay."

But it gets worse. That employer contribution isn't really "matching." It's compensation you already earned that they're withholding. They budget for total employee compensation costs. If they weren't putting money in retirement accounts, that money would have to go somewhere. Some would go to higher salaries, some to better benefits, some to actual cash bonuses.

Instead, they've found a way to pay you with your own future earnings while getting immediate tax deductions and looking generous. You do the work today, generate value today, but can't access part of your compensation until you're too old to use it for anything meaningful like starting a business, buying land, or escaping the job that's slowly killing you.

Meanwhile, if you try to access your own earned compensation early, they hit you with penalties of 10% plus income taxes. So if you're in a 22% tax bracket and take an early withdrawal, you're paying 32% just to access money you already earned through your labor. That's a higher penalty rate than most credit cards charge.

Think about that. They punish you more severely for accessing your own money than banks punish you for borrowing someone else's money.

Why Penalties Don't Matter

Everyone focuses on early withdrawal penalties like they're financial suicide. Let's do some actual math.

Say you have $50,000 in a 401(k) at age 35. If you withdraw it early, you pay a 10% penalty plus income tax, about $16,000 total, leaving you with $34,000.

Sounds terrible, right? But what if you use that $34,000 to eliminate debt or build capabilities that reduce your living expenses by $500 per month? Over 30 years, that's $180,000 in reduced expenses.

Compare that to leaving the money in the account. Even if it doubles every 10 years, that $50,000 becomes $400,000 by age 65. But with 3% annual inflation, that $400,000 has the purchasing power of about $165,000 in today's money.

And that assumes everything goes perfectly: no market crashes, no currency devaluation, no healthcare emergencies that drain the account, no changes to retirement laws.

Here's the kicker. If you're 65 with $400,000 in a retirement account, you still can't buy independence. You're still dependent on grocery stores, utility companies, healthcare systems. You've got money but no capabilities.

If you're 65 with 30 years of experience growing food and living independently, you're infinitely more secure than the person with a fat retirement account and no practical skills.

Breaking Free by Class

The system has everyone gambling at different stakes, but everyone can start reducing their bets. This list is obviously not exhaustive, and you'll need to get creative for your own scenario. What worked for us may not work for you. Here's how to begin, regardless of where you're starting:

The Poor: Stop Feeding the Parasites

If you're living paycheck to paycheck, you're already taking the biggest financial risks. Payday loans, check cashing fees, and rent-to-own schemes are bleeding you dry faster than any early withdrawal penalty ever could.

Actions to take:

- **Eliminate predatory services**: That 400% interest payday loan costs more than any retirement account penalty. Try family help, selling items you don't need, or picking up gig work before borrowing at predatory rates.
- **Learn basic repair skills**: YouTube University can teach you to fix your car, appliances, and clothing. Every repair you do yourself is money that stays in your pocket.
- **Stop renting everything**: Rent-to-own furniture costs 3x retail price. Buy used, fix what breaks, own what you need.

Working Class: Own Your Labor

You're trading hours for dollars while other people profit from your work. The goal is to own more of what you produce and depend less on what others control.

Actions to take:

- **Develop portable skills**: Learn trades that can't be outsourced—plumbing, electrical, mechanical repair. These skills generate income anywhere and reduce your dependence on others.
- **Stop leasing your life**: Car payments, equipment rentals, subscription services. Buy used, maintain it yourself, own it outright.
- **Build side income streams**: Use your existing skills to generate money outside your main job. Even $200/month extra gives you options your employer can't control.

Middle Class: Stop Playing Status Games

You're trapped by lifestyle inflation and the need to look successful. Most of your spending serves other people's profit margins, not your actual needs.

Actions to take:

- **Eliminate status debt**: The expensive car, the bigger house, the luxury vacations you can't afford. Redirect that money toward things that reduce your expenses instead of increasing them.
- **Cut the subscription economy**: Cancel everything that charges you monthly for access to things you don't own. Redirect that money toward buying tools, skills, and capabilities.
- **Downsize strategically**: Move to a smaller house in a rural area where costs are lower and you can grow food, repair things, and live more independently.

Upper Middle Class: Convert Paper to Real Assets

You've accumulated wealth on paper, but it's all dependent on systems you don't control. Time to diversify into things that work regardless of what happens to markets and institutions.

Actions to take:

- **Use home equity wisely**: Instead of leveraging up to buy more real estate, use equity to buy land, solar systems, or businesses that generate income independently of financial markets.
- **Stop chasing paper returns**: Take some money out of stocks and bonds and put it into productive assets. Tools, equipment, skills, and infrastructure that generate value regardless of market conditions.
- **Build location independence**: Create income streams that work from anywhere, so you're not trapped in expensive areas by job requirements.

The Rich: Diversify Out of the System

Your wealth is almost entirely dependent on systems continuing to work the way they have been. But systems change, and when they do, paper wealth disappears while real capabilities remain valuable.

Actions to take:

- **Convert financial assets to productive assets**: Buy farmland, manufacturing equipment, businesses that produce essential goods. Things that generate value regardless of currency or market stability.
- **Develop independence infrastructure**: Solar systems, water sources, food production, backup power. Build systems that work when the grid doesn't.

- **Invest in skills**: Instead of hiring others to do everything, learn to do essential things yourself. When money becomes worthless, capabilities remain valuable.

The pattern is the same at every level: stop depending on systems designed to extract from you, start building capabilities that work regardless of what those systems do. The amount of money involved is different, but the principle is identical.

The Real Risks Nobody Talks About

The "safe" path of leaving money in retirement accounts assumes everything keeps working the way it has been. But what if it doesn't?

- **Healthcare System Collapse:** Your retirement savings won't matter if medical care becomes so expensive that a single serious illness wipes out everything you've saved. Or if the healthcare system simply breaks down under the weight of an aging population and corporate profit extraction.
- **Currency Devaluation:** Inflation has already destroyed the purchasing power of every previous generation's retirement savings. What makes you think your dollars will be worth anything in 30 years when the government keeps printing money to pay for promises they can't keep?
- **Market Manipulation:** The stock market isn't a reflection of economic reality, it's a casino where the house always wins. When it crashes again (not if, when), your retirement account balance disappears while the fund managers still collect their fees.
- **Political Instability:** Retirement accounts depend on stable political systems that honor property rights and enforce contracts. Look around. Does that seem like a safe bet for the next 30 years?

- **Climate Disruption:** Extreme weather events are already destroying infrastructure, flooding coastal cities, and making entire regions uninhabitable. Your retirement portfolio can't buy you security from climate collapse.
- **Technological Displacement:** AI and automation are eliminating entire job categories. What happens to retirement planning when half the workforce is unemployed?

The "responsible" choice isn't to trust these systems for decades. The responsible choice is to build something you can control right now, while you're young enough and healthy enough to do the work.

The Phased Approach That Actually Works

We didn't burn everything down at once. Alexys moved from a toxic InfoSec Director role grinding through her doctorate while managing Multiple Sclerosis, to a fully remote position with better culture, less stress, and less responsibility. She kept income and benefits while we planned my exit from federal hell.

Multiple Sclerosis is already a battle against your own nervous system. Fatigue that feels like drowning, brain fog that steals words mid-sentence, pain that moves through your body like wildfire. But her high-stress leadership role was actively making her sicker. The constant stress triggered flare-ups. The fluorescent-lit office drained what little energy MS hadn't already stolen. Mandatory in-person meetings meant dragging herself through buildings when walking felt like moving through concrete. The always-on culture demanded performance when her body was screaming for rest. Every deadline push became another assault on an immune system already at war with itself.

The beauty of escaping the system is that it doesn't have to be all or nothing and all at once. You can eliminate the worst parts first, then methodically work your way out of the rest.

Getting Alexys out of that toxic environment proved the point. Her health stabilized, her energy returned, and suddenly we had the mental space to plan instead of just survive. That's the phased approach. Eliminate what's killing you first, then use that breathing room to eliminate the rest.

But our path wasn't a direct one. Sometimes it was stair-stepped. Small wins building on each other, climbing toward something better. Other times it felt like a treadmill, running hard but feeling like you're going nowhere, watching the same scenery pass by month after month. We'd pour energy into what seemed like progress, only to find ourselves back where we started, just more exhausted. There were dead ends that looked promising from a distance. Business ideas that seemed perfect until we hit the wall of reality. Opportunities that evaporated the moment we reached for them. Paths that seemed to lead somewhere until they simply... didn't. Each detour taught us something, but it felt like we were solving a maze blindfolded, feeling our way along walls we couldn't see, retracing steps we thought we'd left behind forever.

It's convoluted and confusing and filled with uncertainty until the moment it's not. Very few people have felt that transition, when the fog suddenly lifts and you realize you've made it to the other side.

Here's what real freedom looks like: if safety nets disappeared tomorrow, there'd be no panic, no scrambling, no desperate calculations about mortgage payments and medical bills. No golden handcuffs. No trapped desperation. Just the clean ability to walk away because we built something that doesn't depend on anyone else's continued benevolence.

Other Partial Exit Strategies:

Not everyone can cash out everything and buy land tomorrow. Here are specific strategies based on your situation. Building that freedom requires resources, and most of yours are probably locked away in retirement accounts you can't touch without penalties. But those penalties might be the best investment you ever make. Here's how to think about accessing trapped money to fund your escape:

If You Can Access 50% of Your Funds

Take it. Use that money to build practical wealth: solar panels that eliminate electric bills, tools that let you repair instead of replace, land in a rural area where costs are lower. Leave the other half if you need the psychological security.

$25,000 can buy a solar system that saves $300/month in electric bills, which is $108,000 over 30 years. Or professional training in trades that can't be outsourced. Or eliminate debt that's costing you 18-24% annually in interest.

If You Can Only Access Your Contributions

Most retirement accounts let you withdraw your direct contributions without penalties. Even $15,000 is enough to eliminate credit card debt, buy a reliable vehicle with cash instead of taking payments, or build an emergency fund that doesn't require permission to access.

If You Can't Access Existing Funds

Stop contributing above any employer match. Redirect that money toward things you control immediately: emergency funds, debt elimination, tools and training for practical skills.

If you're contributing $500 per month to retirement, redirect it to build capabilities that reduce your dependence on monthly payments and corporate employment.

Geographic Arbitrage

Use retirement funds strategically to escape high-cost areas. $30,000 after penalties can cover moving costs to a rural area where housing costs 60% less, plus provide cushion while you establish income in the new location. The monthly savings often exceed what you lost to penalties within 2-3 years.

Skills Investment

$20,000 can buy commercial driver training, electrician apprenticeship, or equipment to start a repair business. These investments generate income immediately, unlike retirement accounts that generate nothing for decades.

The key insight: retirement accounts represent a bet that you'll be better off having money when you're old than having capabilities when you're young. For most people, that's a bad bet.

The Urgency of Now

The biggest gamble is waiting. Waiting for the "right time" that never comes. Waiting for permission from systems that profit from your waiting. Waiting until you're 65 to access money you earned when you were 35.

Every year you wait is another year older, another year more dependent on systems you can't control, another year further from the physical capability to build alternatives. The 45-year-old who cashes out retirement accounts and learns to grow food is in a better position than the 65-year-old with a full retirement account who can't change a tire.

We're not saying everyone should become farmers. We're saying everyone should recognize that the conventional path is the biggest gamble of all, betting your entire future on systems that have never worked for the people who depend on them, run by people who profit from your dependence.

The money in your retirement account? You already earned it. The penalties for accessing it? They're designed to keep you trapped, not to protect your future. The "responsibility" of leaving it alone? That's responsibility to everyone except yourself.

The real irresponsible choice is trusting your future to people who see you as a resource to be extracted from rather than a human being to be served. The real gambling is betting that systems designed to fail won't fail you personally.

We didn't throw away our future when we cashed out any accounts. We bought our future back from people who were renting it to us at terms that guaranteed we'd never be able to afford it.

That money was always ours. We just stopped pretending it belonged to someone else.

Chapter 23: The Exit Funnel

We don't manipulate people. Period. We don't use psychological tricks, false scarcity, or any of the bullshit tactics that most online businesses use to pressure people into buying things they don't need with money they don't have for promises that won't be kept.

Our "funnel" isn't a funnel at all. It's just honesty about what we do and how people can engage with it if they want to. No manipulation. No pressure. No artificial deadlines or limited-time offers designed to bypass rational decision-making.

People find us through social media or word-of-mouth. If they want to know more, they can visit our website. If they want updates, they can join our email list. If they want to buy products, they can order them. If they want to learn skills, they can take our courses. If they want personal advice, they can book consultation time.

Every step is optional. Every step provides value on its own. People engage at whatever level makes sense for their situation, goals, and budget. We don't push anyone toward more expensive options or bigger commitments.

The Anti-Funnel That Actually Works

Traditional sales funnels are designed to move people from awareness to purchase through a series of psychological manipulations that overcome objections and create artificial urgency. They treat potential customers like targets to be conquered rather than people with real problems who might benefit from real solutions.

Our approach treats people like intelligent adults who can make their own decisions about their own lives when given honest information about available options. We present what we offer. People decide what, if anything, they want.

The difference is respect. We respect people's autonomy, intelligence, and ability to evaluate whether our products and services serve their actual needs. Traditional funnels assume people are too stupid or too emotional to make rational decisions without manipulation.

We attract people who are already questioning the systems we escaped from. People who are looking for alternatives to conventional approaches that aren't working for them. People who value authenticity over polish and substance over marketing tactics.

These people don't need to be convinced or manipulated. They need honest information about what we've learned and what we offer. They need to know whether our experience and knowledge can help them solve problems they're actually facing.

Natural Progression vs. Forced Conversion

People who are ready for change will take action. People who aren't ready will consume content passively or move on to something else. Both outcomes are fine with us because we're not responsible for other people's choices or timelines.

Someone might follow us for months before buying their first dozen eggs. Someone else might book a consultation call after reading one newsletter. Another person might take our processing course without ever buying physical products. All of these relationships serve different needs and provide different value.

The natural progression from awareness to engagement to purchase happens when people are ready, not when we decide they should be ready. Artificial pressure creates artificial urgency that leads to buyer's remorse and refund requests and negative word-of-mouth from people who felt manipulated into decisions they weren't prepared to make.

We'd rather have fewer customers who are genuinely satisfied than more customers who feel tricked into purchasing things they didn't really want or need. Quality relationships beat quantity metrics every time when your goal is sustainability rather than maximum short-term revenue extraction.

Value at Every Level

Someone who never buys anything from us but applies what they learn from our free content still benefits from the relationship. Someone who only buys eggs occasionally still gets value from knowing where their food comes from. Someone who takes our courses but never books consultation time still gains knowledge they can apply to their own situation.

This isn't charity or a marketing strategy. It's recognition that value exchange doesn't always involve money changing hands. People share our content with others who might benefit from it. They refer customers. They provide feedback that helps us improve what we offer.

The relationship is symbiotic rather than extractive. We provide information, products, and services that help people live more independently. They provide support, feedback, and community that helps us continue doing what we're doing.

Value at every level means we don't bait people with free content designed to manipulate them into expensive purchases. The free content is genuinely useful on its own. The paid products solve different problems or provide more detailed information for people who want deeper engagement.

Transparency Instead of Tactics

We tell people exactly what they're getting before they buy it. No hidden costs. No bait-and-switch. No upsells that turn a reasonable purchase into an expensive commitment they weren't prepared for.

Our product descriptions are honest about what's included and what's not included. Our course content is clearly outlined so people can evaluate whether it addresses their specific needs and knowledge level. Our consultation calls have clear agendas and realistic expectations about what can be accomplished in the time allocated.

We don't promise life-changing transformation or guarantee specific outcomes that depend on factors we can't control. We promise to share what we know honestly and let people apply that knowledge according to their own circumstances and capabilities.

Transparency builds trust with people who value honesty more than hype. These are our ideal customers, people who want to make informed decisions based on accurate information rather than emotional appeals and artificial pressure.

Respect for Decision-Making Autonomy

People are smart enough to evaluate whether our offerings serve their needs without high-pressure sales tactics or psychological manipulation designed to bypass rational consideration. They can assess their own situations, budgets, and priorities better than we can.

Our job is to provide accurate information about what we offer and why it might be useful. Their job is to decide whether our products and services align with their goals and resources. Clean division of responsibility that eliminates manipulation from the relationship.

When people decide not to buy something, we respect that decision without trying to overcome their objections or convince them they're wrong about their own needs and circumstances. Maybe our offering isn't right for them. Maybe it's not the right time. Maybe they found a better solution elsewhere.

All of those outcomes are acceptable because the goal isn't to maximize sales. It's to provide value to people who can benefit from what we offer. Quality relationships with satisfied customers matter more than conversion rates and revenue optimization.

The Long Game vs. Quick Extraction

We're building something sustainable that will support us for decades, not trying to maximize short-term revenue that requires constant customer acquisition to replace people who feel ripped off and never buy again.

Customer lifetime value comes from trust, satisfaction, and ongoing relationships rather than one-time sales to people who get pressured into purchases they regret. People who trust us buy from us repeatedly and refer others. That's more valuable than aggressive tactics that extract maximum revenue from each interaction.

The long game requires patience and consistency rather than urgency and pressure. It means focusing on quality over quantity, satisfaction over conversion rates, relationships over transactions.

Building a sustainable business around authentic relationships takes longer than building revenue through manipulation and pressure tactics, but it creates something that lasts instead of something that burns through customers and requires constant replacement with new targets who haven't learned to avoid high-pressure sales approaches.

Exit Strategy for People Ready to Exit

The people who find value in what we offer are people who are ready to question assumptions about how life has to be lived. People who are looking for alternatives to systems that aren't serving their needs. People who want to reduce their dependence on institutions that extract more value than they provide.

Our content and products serve people who are planning their own exits from systems that are slowly killing them. Not people who want to optimize their participation in those systems or find ways to make them more tolerable.

This clarity about who we serve eliminates the need to convince or convert anyone. The people who are ready will recognize themselves in what we share. The people who aren't ready will find something else that better matches where they are in their journey.

We're not trying to create dissatisfaction with conventional approaches or convince people that our way is the only way. We're sharing what worked for us and providing resources for people who want to try similar approaches adapted to their own circumstances and goals.

The exit funnel serves people who are already looking for the exit signs. We just make those signs more visible and provide maps for people who are ready to walk toward the door.

Part V: The Hard Truths

Chapter 24: Processing Day

The worst part of my routine is processing day. I'm always kind of quiet in the mornings when I have to kill animals. Going through my mind how to make this as quick and painless as possible.

It's not something you get used to. Anyone who says they get used to it is either lying or shouldn't be trusted with living creatures.

I've processed hundreds of animals and no animal acts the same. Each one has its own personality. Their own stress responses. You can't just develop a routine and assume it will work every time.

I've processed chickens, rabbits, and quail. All different methods. Different equipment. Different techniques based on their anatomy and behavior and size. Chickens require different restraint than rabbits. Rabbits are different from quail. Quail are so small that the margin for error is much smaller.

Some chickens are calm and docile. They seem to accept what's happening without struggle. Others may panic. Some rabbits freeze when handled. Others kick and scratch and try to escape.

The responsibility of ending a life weighs on you every time. It should weigh on you every time. The moment it stops being significant is the moment you've lost respect for life.

My quiet morning ritual includes checking equipment. Sharpening knives. Preparing processing areas. Mentally rehearsing the procedures. The animals don't know what's coming. They don't need to know. Their last experience should be as calm and stress-free as possible.

Processing day reminds you that eating meat has costs beyond money. Every meal with animal protein represents a life that ended for your dinner. Industrial meat production hides this reality behind slaughterhouses and packaging. When you raise and process your own animals, you can't avoid it.

The efficiency of the process doesn't make it emotionally easier. It makes it morally necessary. Quick, clean kills reduce suffering. But they don't eliminate the weight of responsibility.

Chapter 25: The Public is Your Enemy

What makes me think "fuck this shit" on a regular basis is my new relationship with the public. People are challenging in ways I didn't anticipate when we started this business.

People want farm experiences without understanding farm realities. They want cute animals and fresh eggs and organic vegetables. They want pastoral aesthetics without comprehending the work and expense and risk that goes into producing what they think they're entitled to buy at industrial prices.

Customers expect us to educate their children about farm life, while selling products at prices that don't reflect the cost of providing educational entertainment. They want personal relationships with farmers while treating farmers like service providers who should be grateful for their business.

The entitlement of people who think supporting local agriculture means getting premium products at discount prices reveals their fundamental misunderstanding of economics. They expect small-scale farmers to subsidize their food costs through unpaid labor and below-cost pricing.

People who romanticize farming don't want to hear about animal deaths. Equipment failures. Weather disasters. Predator attacks. Disease outbreaks. Financial stress. Physical exhaustion. Or any of the realities that make farming one of the most dangerous and economically uncertain occupations.

The public wants farmers to be grateful for their business while treating farming like a hobby rather than a profession. Like entertainment rather than essential work. Like a lifestyle choice rather than a business that requires profitability to continue operating.

Customer interactions that should be simple transactions become negotiations. People expect discounts for bulk purchases. Special accommodations for pickup schedules. Detailed explanations of production methods. Guarantees about product quality. Personal relationships that extend beyond business into friendship territory where boundaries become unclear.

The expectation that farmers should be educators, entertainers, therapists, and friends in addition to producers reveals how disconnected most people have become from food production realities. We're trying to make a living, not make friends.

Chapter 26: Nobody Actually Cares If You Live or Die

What I miss most about my old life is believing that people cared about me. The illusion of social connection and family loyalty that exists as long as your choices conform to other people's expectations. That all disappears quickly when you make decisions that challenge their assumptions.

Having my family disown me is something I think about daily. Not because I regret our choices. Because it reveals how conditional their love and support actually were. Family relationships that seemed solid turned out to be contingent on my continued participation in systems they approved of.

When I stopped seeking their approval and started living according to my own values, they stopped treating me like family. They started treating me like a problem to be managed or ignored.

The realization that no one outside of Alexys and our animals cares if I live or die is a sobering thought. More people should come to grips with this since I'm not unique. Most people live believing they have social safety nets and support systems that would help during emergencies. But these safety nets are often illusions.

Most exist only as long as people conform to social expectations. The moment your choices force other people to reflect on their own lives, they check out.

My family members who disapproved of our lifestyle changes weren't offering alternative solutions to the problems we were trying to solve. They were just expressing disappointment that we weren't continuing to live in ways that made them comfortable with their own choices to remain trapped.

Social isolation from making unconventional choices reveals the difference between relationships based on genuine care and relationships based on shared participation in systems that create artificial bonds. Bonds through common experiences of suffering and compliance with social expectations.

The loneliness of opting out is real. But it's cleaner and more honest than fake connection from participating in systems that drain your energy and compromise your values while providing the illusion of belonging.

We never excluded anyone. They excluded themselves. Our independence made them uncomfortable, and instead of facing that discomfort, they walked away. If this happens to you, let them.

The only mistake we made was not doing this sooner.

Chapter 27: Quail Are Assholes But They Work

The chore I hate most is cleaning quail hutches. Quail are a pain in the ass. But they're efficient animals, and I deal with it because they quickly produce so many eggs and so much meat.

At just six weeks old, Coturnix quail are sexually mature, lay eggs, provide fertilized hatching eggs, and are ready to process for meat. Compare that to chickens, which take about six months to lay or grow out to processing weight. Unless you get meat broiler birds, which require way more space, will never lay eggs, and serve no real purpose aside from meat. Chicken are hardly efficient animals when compared to quail. And you can't compare the taste of any chicken to that of quail. Just ask Alexys. When it comes to poultry meat flavor, quail are gods.

The efficiency doesn't make them pleasant to work with. They're flighty. Nervous. Prone to panic responses that can injure them, kill them, or damage equipment. They produce more waste per pound of body weight than larger birds. They require more frequent cleaning.

Quail hutches need frequent cleaning because waste builds up quickly in small spaces. The smell becomes overwhelming if you wait too long. The cleaning process involves a bunch of shit I won't bore you with, but it includes more than just keeping the birds safe and from escaping. You're emptying and washing water and feed containers. Removing soiled bedding and waste. Scrubbing surfaces. Disinfecting. Replacing bedding and sand. Topping off resources. Then making sure every bird is safely locked in, with safety bars secured.

Their behavior during cleaning can make the job worse. They can panic. They can bolt for any open space. They can injure themselves flying into walls when startled. They can make simple tasks take longer than they should through natural fear responses to anything unfamiliar.

But I'm calm around them. Our birds trust me. I handle them with care. They come to the hutch door when I show up with food. I pet them, I talk to them, and I love them. Every last one, even the assholes.

Despite their annoying behavior and high maintenance needs, quail economics make them worthwhile. Rapid reproduction. Efficient feed-to-egg and feed-to-meat conversion. Small space requirements. Premium market prices for both eggs and meat. The benefits outweigh the bullshit when you measure by productivity and profit instead of personal preference.

My hate-tolerate relationship with quail reflects the kind of compromise working farmers make all the time. You don't have to like every animal you raise. You just have to understand what they're good at and decide if it's worth the trade. For us, it is.

Chapter 28: When Things Die and It's Your Fault

Animals can die on any farm where people are learning through hands-on experience instead of just theory. Not usually the grown animals. But some chicks don't hatch completely or never make it out of the shell.

You wonder if you could have done something different to improve their chances. Different humidity levels. Temperature adjustments. Turning schedules. Incubator positioning. All the variables that impact hatch rates but aren't obvious when you're still learning.

Hatching failures feel personal, even when they're probably just statistical noise. That one chick that struggles and dies half-emerged becomes the symbol of every mistake you think you might have made. It feels like a failure of your care, even if it wasn't.

The urge to intervene when chicks are struggling creates real dilemmas. When does stepping in help, and when does it interfere with natural selection? Helping weak chicks might save them in the moment but pass on traits that make future generations weaker. Doing nothing feels like giving up.

Other than hatching failures, we've been lucky enough to avoid accidental deaths caused by our negligence. That's partly because we're millennials who over-plan and overthink every fucking aspect of animal care. We research like lunatics. Build systems with backups. Keep logs. Follow preventative health routines. We monitor everything.

We probably go overboard. But that obsession with getting it right helps prevent the kinds of deaths that come from carelessness, missed details, or not thinking ahead.

Overthinking animal care creates anxiety about problems that might never happen. But it also helps stop the ones that actually do. It's a tradeoff.

Losing an animal to your own mistake hits different than processing one for meat. Processing is the reason the animal was raised. It has a purpose. Accidental death is just loss. Loss of life, loss of time, loss of resources, with no nutrition, no income, and no lessons unless you let it hurt and change how you work.

Learning animal husbandry always involves loss. But that doesn't mean loss has to be frequent or careless. You can't eliminate it completely, but you can reduce it with preparation and humility. You can learn from others so your mistakes don't have to cost as much as theirs did.

Chapter 29: Money Pays for Nothing

This lifestyle costs a great deal of time and physical exertion. Very little money is required aside from feeding animals, building their housing (mostly just the initial cost), and our utilities. No HOA. No zoning restrictions. Cheap, simple permitting process. We live in a right-to-farm state. Look it up if you've never heard of it.

The financial reality is that money requirements are much lower than urban or suburban living. But time and energy requirements are much higher. You're doing for yourself what most people pay others to do.

I'm not going to talk about how much we make. Our income is our business. It's not relevant to whether this lifestyle is financially viable. Viability depends more on what you spend than what you earn. More on your skills and resources than your revenue.

We put in a lot of effort, and it's not wasted. That effort turns into food, independence, stability, peace of mind. Things you can't buy from people who sell convenience and illusion.

We used to be genuinely worried about money, back when we had more of it. High income doesn't mean security when you're swimming in debt and dependent on employers who can cut you off anytime they want.

Now we know we can feed ourselves whether the market crashes or a job disappears. That's not just peace of mind. That's power. Food you grow is food no one can take from you. That security isn't theoretical. It's daily reality.

To quote one of our favorite movies, "money buys nothing." Money buys nothing that really matters for survival or wellbeing. Money buys convenience and status and entertainment and access to systems that mostly exist to extract money from people who think they need what they could provide for themselves. The quote is from the movie *Wanderlust*. If you haven't seen it, you're wasting your life. But don't stream it. Buy it. A physical copy.

Shifting from high-income, high-expense living to low-income, low-expense living exposed how much of our past spending was waste. Money spent to impress people we didn't like or to save time we just used to recover from the jobs we had to work to earn the money in the first place.

Real financial independence means needing less. Not earning more. Cutting expenses until your income no longer controls your life. Until your money serves you instead of feeding systems that make you feel like you never have enough.

The psychological relief of not needing much is worth more than the illusion of financial success built on fragile systems and pointless obligations. When your life costs less to maintain, you no longer owe it to people who never cared about your freedom.

Chapter 30: Medical Realities in a Collapsing World

Alexys depends on infused medication to stop her MS progression. Not just to feel better, to survive. Without it, another lesion could form that steals her ability to walk, or critically think, or swallow. The ability to purchase and access healthcare isn't a preference for us. It's a necessity. We don't care what it costs, how many hoops we have to jump through, or how many systems we have to drag ourselves through to get what she needs. The only unacceptable and fearful outcome is losing her to a system that collapses before we're ready.

That's our biggest vulnerability. We can grow food, produce energy, purify water, build shelter, and protect ourselves, but we can't manufacture prescription drugs. We can't perform surgeries. We can't build an MRI in our garage.

We are forced to maintain health insurance even though we hate the systems that create it. We keep working relationships with medical professionals who understand where we're coming from. We make contingency plans for when the official channels become too broken or too expensive to matter.

But we're not delusional. When the healthcare system breaks, people like Alexys are the first to feel it. Chronic illness becomes a death sentence not because it has to be, but because the care stops coming.

This tension lives at the heart of everything we've built. We can opt out of mortgages and corporate jobs and subscription services, but we can't opt out of biology. The system still has hooks in us through her medical needs, and that keeps us tethered in ways that complicate the clean narrative of total independence.

We've made peace with this contradiction because the alternative, gambling her life on ideological purity, isn't an option. We'll work within broken systems to keep her alive and feeling good while building alternatives where we can. That's not hypocrisy. That's strategy.

The medical dependency also shapes our timeline differently than it might for others. We can't just disappear into the mountains and go completely off-grid. If your body has privileged you with this opportunity, take it. We need to stay within driving distance of medical facilities. We need to maintain enough traditional income to cover health insurance and treatments. We need backup plans that account for medical emergencies.

This doesn't invalidate what we've built, it just makes it more nuanced. Independence isn't binary. It's not all-or-nothing. It's about reducing dependencies where you can while managing the ones you can't eliminate.

Chapter 31: Building Real Security

You can't build resilience on rented infrastructure. If someone else owns the power, the pipes, the networks, or the fuel, then they own your limits too. That's not doom prep. That's just the truth. So we started asking what breaks first. What fails fastest when pressure hits. What systems we can't afford to lose.

Energy Independence as Insurance

Fuel won't matter much in the future, not to us. We're building toward full solar capability, piece by piece. Not because it's trendy or greenwashed or grant-funded. Because it works without the grid.

The grid is a dinosaur, held together with bailouts and duct tape. It gets weaker every year while demand rises. And the companies who run it don't care. They'll raise your rates until you're drowning, then call it infrastructure investment.

Depending on solar energy isn't perfect. Panels need replacement. Batteries wear out. The parts come from factories that could vanish overnight. But it's still more reliable than a system designed to fail gracefully for shareholders.

The transition to solar isn't just about energy independence, it's about breaking free from the monthly electric bill that keeps you tied to the cash economy. When your power comes from the sun, you're not vulnerable to rate increases, service interruptions, or corporate policy changes that prioritize profit over reliability.

We started small. A few panels to run essential systems. Basic battery backup for critical loads. Gradually expanding capacity as we learned what our actual energy needs were versus what the utility company told us we needed.

The learning curve was steep. Solar isn't as simple as the sales brochures make it seem. You need to understand battery chemistry, charge controllers, inverters, system sizing, and electrical safety. But every kilowatt-hour we generate ourselves is one less we depend on others to provide.

The psychological impact is as significant as the practical benefits. When the power goes out in the neighborhood, our lights stay on. When energy prices spike, our costs stay flat. When the grid fails, we keep running.

We'll keep a fuel reserve, just enough for machines and emergencies, but we're not building a future that depends on diesel trucks showing up on time. Fuel dependency is just another form of systemic vulnerability. The less we need, the more resilient we become.

Food Security in an Uncertain World

We don't fear food shortages. Our meat doesn't come from factories. Our produce doesn't rely on refrigerated shipping containers. Our food grows twenty feet from the door. If we want quail for dinner, I walk out to the hutch, grab a couple live birds, process them over a bucket in about 10 minutes with a pair of kitchen shears, and once they are washed and brought in the house, we are dining on quail 20 minutes after that.

The difference between food security and food dependency is the difference between capability and consumption. Most people have food security as long as they have money and the supply chains function. We have food security because we can produce what we need regardless of external conditions.

This didn't happen overnight. Building food production capacity takes years of learning, infrastructure development, and system optimization. We started with rabbits because they're efficient and relatively simple. Expanded to quail for efficiency, specialty products, and market diversity. Added chickens for eggs and different flavors of meat.

Each species taught us different lessons about animal husbandry, processing, feed efficiency, and market demand. Each failure, and there were plenty, improved our understanding of what works and what doesn't in our specific environment.

We don't need dairy. It isn't essential to anything we are trying to accomplish, but we do enjoy milk and cookies as well as cheese on our meals. The idea that dairy is essential is a marketing campaign, not a biological truth. Dairy animals require more infrastructure, more expertise, and more time than we're willing to invest for products we can live without.

Vegetable production follows the same philosophy. We grow what we eat and eat what we grow. No exotic crops that require special conditions or inputs we can't provide ourselves. No monoculture that leaves us vulnerable to single-point failures.

Rabbits eat hay, weeds, grass, and scraps. They turn garbage into protein. They don't need high-tech feed or climate-controlled barns. Chickens and quail are the same, hardy, small, efficient. Industrial meat is soft. Ours is not.

We save seeds. We preserve food without power. We know what worked before grocery stores existed. That's the bar. That's what we trust.

The food system we've built isn't just about calories and nutrition, it's about knowledge and capability. Every season we learn more about soil management, pest control, harvest timing, and storage techniques. Skills that can't be taken away by economic disruption or supply chain failures.

Financial Resilience vs. Paper Promises

Bank failures don't scare us. We don't owe anybody.

Our cash is spread. Our assets are real. Skills, tools, relationships, infrastructure. Things you can't freeze with a policy or delete with a software update.

Digital money is a light switch someone else controls. When it turns off, most people starve while watching their screens say "error." But you can't repo a garden. You can't audit a pantry. You can't sanction someone who knows how to slaughter, seed, preserve, and build.

This is how we store value: in work. In knowledge. In things that hold up when the markets don't.

The transition from financial assets to real assets was gradual but deliberate. Every dollar that came out of retirement accounts went into tools, infrastructure, or skills that increase our productive capacity. Money spent on land, equipment, and education provides returns that don't depend on market performance or institutional stability.

We keep enough cash for operating expenses and unexpected costs, but we don't accumulate money for its own sake. Money sitting in accounts is money that's not working for our independence. Money invested in markets is money we don't control.

The goal isn't to get rich, it's to get secure. Security through capability, not capital. Security through what we can do, not what we can buy.

This approach requires a fundamental shift in thinking about wealth and security. Instead of asking "how much money do I need?" we ask "what capabilities do I need?" Instead of optimizing for maximum returns, we optimize for maximum resilience.

The math works differently when you're not trying to maximize profit. A solar system that pays for itself in ten years might not be the best financial investment compared to stock market returns, but it provides energy independence that has value beyond monetary calculation.

Same with food production. Growing your own vegetables might cost more per pound than buying them from the store, but the security of knowing you can feed yourself regardless of external conditions is worth the premium.

Community vs. Independence

We love the people who remain in our lives. But we don't depend on them.

Because people die. They move. They lose their jobs. They get sick. They flake. They change their minds. And none of that makes them bad people, it just makes them human.

We've lost too much to hinge our survival on someone else's availability. When we have extra, we share. When others need help, we help. But the help we give doesn't come from desperation, it comes from margin. We don't owe. We choose.

That's what community should be: voluntary, strong, and rooted in respect, not survival tied to strings and favors and expectations.

The independence we've built makes us better community members, not worse ones. When you're not desperate, you can afford to be generous. When you're not dependent, your relationships can be based on genuine care rather than mutual need.

This distinction matters. Mutual aid from a position of strength is different from mutual dependence from a position of weakness. One builds community. The other creates resentment and obligation.

We maintain relationships with neighbors who share similar values and practical skills. People who understand that self-reliance isn't selfishness, it's the foundation that makes genuine community possible.

But we don't participate in systems that require constant coordination and consensus. We don't join organizations that demand our time and energy for causes we don't control. We don't commit to obligations that compromise our independence for abstract social benefits.

The community we choose is small, local, and voluntary. People who know our names and understand our values. People who respect our choices even when they make different ones. People who would help in a crisis but don't expect constant engagement during normal times.

This isn't isolation, it's selective engagement. Choosing quality over quantity in relationships, depth over breadth in commitments.

What Winning Actually Looks Like

Winning looks like waking up every day and choosing how to spend our time based on our values and priorities instead of other people's demands and expectations. It looks like eating food we raised and trusting where it came from. It looks like working hard on things that matter to us instead of working hard to make other people wealthy.

But winning also looks like the small, daily satisfactions that most people never experience. The sound of animals eating their morning feed. The weight of eggs still warm from the nest. The exhaustion that comes from productive work instead of busy work.

Winning looks like sleeping through the night without rehearsing tomorrow's meetings because there are no meetings. It looks like checking the weather because it affects your work, not checking your phone because you're bored with your life.

It looks like having conversations with Alexys about what we want to do next week, next season, next year, conversations about possibilities rather than obligations. About opportunities rather than deadlines.

Winning looks like being excited about our work instead of enduring it. Looking forward to challenges instead of dreading them. Solving problems that matter instead of managing problems that exist primarily to justify other people's salaries.

Most days, winning looks like routine. Feed the animals, tend the garden, maintain the infrastructure, serve the customers, plan for tomorrow. Simple work that produces visible results and serves clear purposes.

But some days, winning looks like sitting under our walnut tree with the dogs, watching the animals be animals, eating lunch we grew ourselves, and realizing we don't want to be anywhere else or doing anything else.

That's not a feeling we ever had in our previous life. No matter how good the job, how nice the house, how impressive the income, there was always somewhere else we'd rather be. Always something missing. Always the sense that we were delaying real life to serve other people's definitions of success.

Now real life is what we're living. Not what we're working toward. Not what we'll do someday when we retire. Not what we'll enjoy if we ever get ahead of the bills and the obligations.

Success is sustainability, not growth. It's maintaining what we've built without burning out or compromising the values that guided us here. It's being able to say no to opportunities that don't align with our goals, even when those opportunities look good to other people.

The Hard Truths About This Path

As cliche as it sounds, this life isn't for everyone, and that's okay. It requires sacrifices that most people aren't willing to make. It demands skills that most people don't want to learn. It involves risks that most people won't take.

You have to be willing to lose and stop chasing money to gain freedom. You have to be willing to work harder physically and to work less mentally. You have to be willing to be misunderstood by people whose opinions used to matter to you.

You have to be willing to fail at things that matter instead of succeeding at things that don't. You have to be willing to take responsibility for outcomes instead of having other people to blame when things go wrong.

The path requires a fundamental shift in values, from seeking approval to seeking authenticity, from maximizing income to minimizing dependence, from growing assets to building capabilities.

It's not a get-rich scheme or a lifestyle hack or a way to work less while earning more. It's a complete reorganization of priorities that puts independence above comfort, meaning above money, and long-term security above short-term convenience.

Most people who try this path will quit. They'll decide the trade-offs aren't worth it. They'll miss the conveniences they gave up more than they value the independence they gained. They'll go back to their previous lives with new appreciation for things they used to take for granted.

That's not failure, that's learning. Everyone should understand what they're actually choosing when they accept the conventional path. Most people have never seriously considered alternatives, so they don't know what they're trading away for security and convenience.

The System's Response to Exit

The system doesn't want you to leave. It's designed to make exit difficult, expensive, and socially unacceptable. Every institution has mechanisms to punish defection and reward compliance.

Financial systems make it costly to access your own money early. Social systems make independence look selfish or irresponsible. Family systems make unconventional choices feel like betrayal.

The more people who exit successfully, the more obvious it becomes that the system serves itself more than the people trapped within it. Success stories like ours threaten the narrative that there's no alternative, that everyone has to play by the same rules, that security requires surrender of autonomy.

Expect resistance. Expect criticism. Expect people to root for your failure because your success makes their acceptance of the status quo look like a choice rather than a necessity.

The system's response will evolve as more people choose exit strategies. Regulations will make self-sufficiency harder. Zoning laws will restrict food production. Licensing requirements will limit home-based businesses. Tax policies will penalize independence.

But the harder they make it to leave, the more obvious it becomes that leaving is the right choice. The desperation of institutions trying to prevent exit reveals how much they depend on captive participants.

If We Win, We Get to Keep Living

If we win, we get to keep living the life we designed instead of the life we were sold. We get to stay home, stay together, and stay focused on what actually matters to us. Everything else is other people's problems to solve or ignore as they choose.

Winning means we never have to go back. Never have to return to jobs that slowly kill us, houses that feel like prisons, relationships that exist primarily to serve other people's comfort with their own choices.

Winning means we've built something sustainable enough to last as long as we do. Something that doesn't require constant growth or optimization or external validation to remain viable.

The ultimate victory is proving to ourselves that alternatives exist. That the conventional path isn't the only path. That security can come from capability rather than compliance.

We don't need to convince anyone else. We don't need to build a movement or change the system or save the world. We just need to demonstrate that it's possible to step outside systems that feel inescapable and build something better with your own hands.

If enough people do that, the system changes itself. If not enough people do it, at least some people escaped. Either outcome is acceptable.

The choice, your choice, has always been yours. Even when they convinced you it wasn't.

Chapter 32: The Work That Stays

We didn't come here chasing a dream. We came because we were done.

Done with the performance. Done with the debt. Done with the long, slow bleed of meaning.

There's this fantasy people like to project onto folks like us, move to the country, grow your own food, breathe clean air, and call it peace. But what they don't see is the collapse that came first. The stripping away. The failure of everything we were taught to believe would keep us safe.

Before the land, there was grief. Before the garden, there was hunger. Before any of this could become real, we had to let everything false die.

No one tells you how long it takes to come back to life after a slow death. How heavy the silence is when you leave behind a world built on noise. You think you'll be relieved. You think the stress will vanish once you cut the cords. But what rushes in is the truth, and it doesn't always feel good.

You lose people. People who can't understand why you'd walk away from a life they're still trying to survive. People who feel like your exit is an insult to their endurance.

You lose ease. You trade comfort for control. Hot showers and takeout for callused hands and the kind of hunger that makes dinner taste holy.

You lose illusion. No one's coming to save you out here. That sounds poetic until the roof leaks, or water runs dry, or you lose the one animal you couldn't afford to lose.

And still, somehow, this was better.

Because out here, everything costs, but nothing is wasted.

You can feel the weight of every choice, but also the worth.

You eat what you grow. You waste less. You give more. You finally sleep at night, not because the world is fixed, but because you're not lying to yourself anymore.

No one claps when the seeds sprout. No one gives you a bonus for harvesting before the frost. But the work still matters.

The hands still learn. The body still remembers.

And slowly, over time, the panic fades. The quiet starts to feel less like absence and more like presence. The soil starts giving back what you gave to it. You don't rise every morning in a panic to make someone else rich. You rise because something needs to be fed. Something that gives back.

You stop needing the world to notice. You stop asking for permission to stay gone.

You start to belong to something again, something older than the noise, deeper than the culture, and more honest than the praise.

You start to see that this is what stays.

Not the job title. Not the followers. Not the house or the leather seats or the endless emails that eat your life.

Just this.

This quiet life built with brutal honesty and care. This handmade safety. This return to real work, done with real hands, for real reasons.

This is what remains when everything else burns. This is what we built when we stopped waiting for permission. This is what winning looks like when you define victory for yourself.

The system will collapse. Systems always do. When it happens, we'll still be here. Still growing food. Still caring for animals. Still choosing our own work. Still living our own lives.

The collapse won't touch us, because we've already collapsed everything that needed collapsing. We've already rebuilt everything worth rebuilding. We've already learned everything worth learning.

When the system collapses, we don't have to. We'll just keep living.

That's the point. That's the plan. That's the promise we made to ourselves when we decided to stop dying slowly and start living deliberately.

This is what's possible when you refuse to accept what you're told is inevitable. This is what happens when you build from the ground up instead of hoping someone else will fix what's broken. This is the life that's waiting on the other side of fear.

The only question is whether you want it badly enough to walk away from everything that's keeping you from it.

The exit signs are always there. They're just not always marked clearly. And they're definitely not easy to see from inside the cage.

But once you see them, you can't unsee them. Once you know there's another way, staying becomes a choice. And once it's a choice, everything changes.

The door was never locked. We just forgot how to turn the handle.

Now we remember.

We remember that every small moment is carved over seasons, slowly, painfully, and without applause.

The End

www.ingramcontent.com/pod-product-compliance
Lightning Source LLC
Chambersburg PA
CBHW031508120626
46545CB00005B/1787